THE FORGOTTEN EXPEDITION

THE 1907 BAKER UNIVERSITY BIOLOGICAL EXPEDITION INTO IDAHO AND OREGON

by

Leon R. Powers

Look-About Books Nampa, Idaho

Look-About Books
Box 1907
Nampa, Idaho 83653

Visit our Website: www. leonpowers.com

First Edition

Library of Congress Control Number: 2010905820

Powers, Leon R (Leon Ray)
The Forgotten Expedition: the 1907 Baker University biological expedition into Idaho and Oregon/Leon R. Powers.—1st. ed.

ISBN 978-0-9800208-1-6

Natural history – Idaho history – Ornithology – Idaho biological expedition (1907) – Baker University – Elmer Perry Monahan I. Title

Cover art by Peter A. Gaede; Cover design by Paul Kinsman

Illustrations by Leon R. Powers and Paul Kinsman

Printed in the United States of America

DEDICATION

To Elmer Perry Monahan Jr., whose
presence, friendship and encouragement
made the telling of his father's story
all the more imperative and fulfilling.

CONTENTS

PREFACE

No one remembers the Baker Expedition of 1907. Of course, it is difficult to recall something that went largely unnoticed when it took place. And such was the fate of this trek. The sponsor of the expedition, Baker University, has no present-day memory of the event and in fact, has precious little archival proof that it ever happened. The repository of the expedition's mother lode of scientific specimens (birds, mammals, herptiles, insects, and plants), the once elegant Baker University Museum of Natural History, long ago disintegrated into shambles until its ultimate demolition in 1976. Along with that expunging of dusty specimens went the tangible evidence of the 1907 trek. If there ever was a museum catalogue or expedition report detailing those specimens collected from Idaho in that summer of 1907, they too have vanished. The expedition leader, Elmer Perry Monahan, never published nor otherwise provided a summary of his faunal and floral collection to Idaho wildlife agencies. Until his diary was accidently discovered in 1966 and published in 1971, even the Idaho history pundits had no documentary proof of our state's last pioneer scientific expedition.

Monahan wasn't a flashy or pretentious person. He didn't even have his name inscribed in his diary. The 1907 Baker University expedition was a pioneer adventure on wagon wheels powered by a team of horses and led by the curiosity of two unknown naturalists, Elmer P. Monahan and Will Green. It may as well have been a phantom expedition for no more attention than it accrued. Its hundreds of collected animal and plant specimens themselves remain specters of lost data.

Over a half century after the fact, the publication of Monahan's diary finally broke the news of this 1907 biological expedition to the select few readers of *Idaho Yesterdays,* the journal of the Idaho Historical Society. Yet little scientific content survives in the diary account, and the smidgen that does remains effectively hidden since biologists rarely poke their noses into history journals when sniffing out scientific data. But I am a native son of Idaho, intrigued by both state history and natural history. I found and read the published Monahan diary in 1973, and since then have been unable to put that historical trek out of my mind.

The diary account, though enlightening, falls far short of recording the expedition's complete journey. The beginning 40 days of their Idaho stay, from May 12 to June 22, go unrecorded, as do the final 44 days in the fall, from September 18 through November 1st. The diary highlights much of the expedition's happenings from June 22 through September 17, 1907 but even here it tantalizingly ignores the fundamental question of the identities of the hundreds of collected specimens. As a biologist and naturalist, that was the burning question from my first reading of the diary – exactly *what all* did they collect in Idaho that summer of 1907? The diary mentions only a few specific names of animals taken or observed along the way. Beyond that, only occasional and typically very general remarks are written about wildlife encounters. The brief and rather prosaic diary narrative begs the question: could Monahan also have recorded the technical aspects of the expedition elsewhere in a field journal? That would have been a rather logical log for Monahan, the taxidermy professor and curator of the Baker University Museum of Natural History. With that possibility could such a field journal still exist? This question long plagued me.

Between 1973 and 1976, and again in 1991, I corresponded with various Baker University biology professors, school administrators and archivists inquiring about any museum specimens, museum catalogues or any other documentation of Monahan's 1907 Idaho trek. Despite their cooperation and earnest efforts, the search was fruitless; "hopeless" in the words of one archivist. That quest produced only six surviving bird specimens and the dream of someday, somehow rediscovering the missing inventory of the phantom collection of the 1907 Baker University Expedition into Idaho.

By spring of 2008, I had long given up hope of ever knowing the make-up of the specimen treasure trove of Monahan's 1907 trek. But even still, I couldn't suppress the compelling urge to put the story of that trek into a book. I wanted to retell this intriguing saga for a broader readership, in an expanded version that would include scientific and historical commentary. Simply stated, I *had* to write a book about that forgotten trek, despite its vexingly truant scientific content. While on a three-month hiatus from writing, awaiting the printing of my latest book, *Dead Owls Flying,* I was conflicted about which book to start next. I was anxious to begin writing about the summers I had spent studying Flammulated Owls in the Sublett Mountains of southeast Idaho. But the urgency of the 1907 Expedition book still hung heavy on my mind. How could I decide? For me writing is a slow, catch-as-catch-can process, and I needed to choose and get started soon on one or the other of the books.

At that time, I had just made application for a post office box number to serve as the business address for my Look-About Books self-publishing entity. When the randomly assigned P.O. Box number was presented to me, I received an unexpected nudge in my decision making. My box number turned out to be . . . 1907! I took that as a sign, and this writing began in earnest and with a clearer sense of mission.

The central intent of this book is to fill in the yawning gaps in the existing story of Monahan's 1907 expedition – expanding far beyond what is reported by his published diary. Roughly half of the 1907 odyssey omitted by the diary is now filled in, including virtually all of the century-long missing scientific data. Woven into this new and expanded expose' is commentary on the scientific and historical significance of that expedition.

To accurately decipher and preserve the identities of the recorded specimen names in the field journals I had to reference two early-1900's bird books which I believe Monahan used on his expedition. Though he may have used other sources as well, the two primary ornithological books guiding Monahan's bird studies of that

1907 expedition were: Elliott Coues' 1903 two volume *Key to North American Birds,* and Chester Reed's 1906 two volume *Bird Guide* (full citations later in this story where I discover reference to them in the field journals).

Although this book reports, for the first time ever, much of the scientific acquisition of that pioneer expedition, it is not intended to be a purely scientific treatise. It is about more than Idaho history and natural history. In addition to its technical import I attempt to capture and portray both the adventure of Monahan's experience and the enchantment of place, set on the Idaho landscape of 103 years ago. Through this portal of time, rumbling along with Monahan and Green in their horse-drawn wagon, we see afresh both the rawness and grandeur of early Idaho.

Purposely, I have enriched the storyline with present-day observations of wildlife recorded along Monahan's old trail. This book provides a helpful guide to anyone who cares to retrace, within these pages or over the actual landscape, the fascinating story of the 1907 Monahan-led, Baker University Expedition.

This book brings back to life the characters, the adventures and the scientific bounty of the 1907 Baker University Expedition, which up until now, have long remained forgotten.

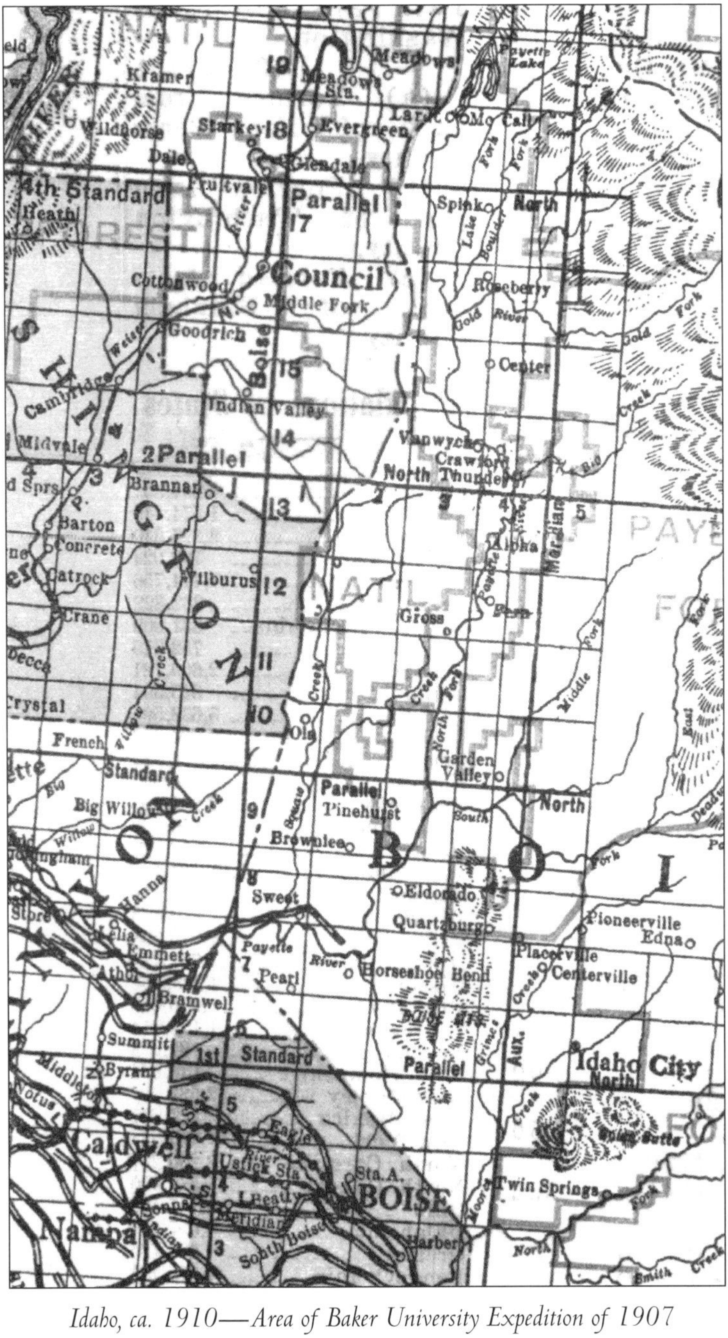

Idaho, ca. 1910—Area of Baker University Expedition of 1907

PART I

DISCOVERING THE FRAGMENTS

·1·

A DIARY FROM NOWHERE (1966)

Kansas State University history professor, Dr. Homer Socolofsky, gingerly opened and began thumbing through the weathered little notebook that had just arrived at his mailbox. He was anticipating a newly discovered diary of the venerable Kansas political figure, Arthur Capper, former publisher, governor and long time U. S. Senator. Four years earlier, in 1962, Socolofsky had published the only extensive biography of this early 1900's politician. For that reason this freshly exhumed, half-century-old diary had been directly forwarded to him, even though it originally surfaced elsewhere, in the Special Collections Office at a rival institution, the University of Kansas. It took but a few moments of perusal by the professor to realize that the diary he held was not penned by Arthur Capper, or by anyone else he knew for that matter. Yet, the substance of the diary held its own bit of historical intrigue. As he read further he discovered that the document was actually a daily record of a hither-to-fore unreported 1907 scientific expedition into central Idaho. A journey far removed from Kansas, yet, it retained strong Kansas ties through its trekkers, a team of biologists from Baker University, a small Methodist college in Baldwin City, Kansas. The expedition's leader and principal diary keeper was Elmer Perry Monahan, a 1905 graduate of the institution and for several years thereafter, one of their biology professors.

Not far into the diary, Socolofsky also uncovered the reason for its erroneous link to Arthur Capper. Curiously, and for reasons known only to them, the two biologists on the 1907 trek assumed aliases of contemporary Kansas political figures. The leader, Monahan, took on the alias of "Cot" (James T.) Pringle, a 1903 – 1907 controversial Kansas State Representative. It was Monahan's expedition partner, William D. Green, a 1907 graduate of Baker, who took on the alias of Arthur Capper. Presumably, these many years later, the discoverers of the diary first noted Arthur Capper's name in the account and misconstrued the diary as being penned by him.

No doubt a little disappointed in the misidentified diary, Socolofsky nonetheless recognized the historical significance of the 1907 account, not only to the annals of Baker University, but also to the folks out west in the area traversed by the expedition. Because of its Idaho setting, Socolofsky contacted one of that state's prominent historians, Merle Wells, in September of 1966 to see if the Idaho Historical Society might be interested in this find. Wells responded with immediate interest in seeing the diary published, so Socolofsky began its preparation. The following month he contacted Baker University personnel, informing them of the recovered diary, and confirmed the records of the old expedition and the identity of its participants. Over the next several months, Socolofsky made several fact-finding visits to the Baker campus and carried on a fairly extensive correspondence with campus officials. By early 1969 he had the diary deciphered and ready for printing in *Idaho Yesterdays,* the journal of the Idaho Historical Society. After some inexplicable printing delays (an all too common bane of publishing on meager budgets) the 1907 diary account finally appeared in the Winter, 1970-71 issue of the Idaho historical journal.

It was barely two years later, in the fall of 1973, that I stumbled onto Socolofsky's report of the newly discovered Idaho expedition. Then a graduate student at Idaho State University, I was writing a paper for an Idaho History class. The topic of my paper was early Idaho biological expeditions. Of course, the Lewis & Clark Expedition was the first and foremost of those explorations, followed by a smattering of other, lesser treks that also cataloged the flora and fauna within our state. Most of these minor expeditions were reported old documents of government-sponsored railroad surveys, or geological expeditions of the 1800's. Although the quest for mineral wealth and transcontinental passageways spurred most of these searches afield, several of the surveys enlisted naturalists as well as geologists. With my interest in field biology, I wanted to know the particular Idaho wildlife species recorded by those pioneer naturalists as they trekked through our state.

I still recall vividly that day when I stood between the tall book shelves in the Idaho State University library thumbing through recent issues of *Idaho Yesterdays.* It was my first introduction to the Idaho Historical Society publication. I was curiously perusing articles, mainly acquainting myself with the journal when I flipped a page and spotted the title, "A Scientific Expedition to Idaho Mountains, 1907". For a moment I just stared blankly. Then, I realized that I was looking at a freshly minted account of a hither-to-fore unknown Idaho expedition; a brand new expose', quite apart from the dusty government reports of old. Not only was it a perfect capstone account for my research paper on early Idaho biological expeditions, it also dawned on me that I was probably the only biologist in the country who – being a quasi-history buff – was aware of this new report. This story offered me a fresh, time-warp look into Idaho's wildlife past. I could hardly wait to read the account and see what fauna and flora had been observed and collected on this just-reported scientific trek. I practically tripped over myself heading for the checkout desk.

The journal article was even more fascinating than I'd hoped. Reading Monahan's account of the old expedition's trek from their initial encampment on Robie Creek, to their serpentine passage over rough mountain roads through Idaho mining country, and finally into Long Valley – all in a creaky old horse-drawn wagon – utterly captivated me. My naturalist spirit settled vicariously into their company over every dusty mile. As I read each day's events, I strained to see the wild 1907 Idaho landscape through their eyes. It was like peering over their shoulders at each bird, insect, plant and other wildling they observed or collected each day. With them I traipsed pioneer trails and rutted roads. I traveled back in time with two kindred spirits who let me explore with them the rugged, brush-choked canyons and forest glades. I gazed through their searching eyes at the vagaries of Idaho's wildlife – a century past. With them I smelled the smoke of "mosquito smudges" and many campfires, and listened to the nighttime calling of wolves and owls from distant ridgelines. I enjoyed a spiritual camaraderie with Monahan and Green akin to what I've known in real life with my own student biologists during our many field studies together. I dreaded the diary's finale. I didn't want it to end. I didn't want to terminate my wild and captivating summer with these pioneer naturalists in the heart of Idaho's mountains, vicarious as it was.

I confess that when I read the final page of the 1907 story, I sat for several minutes with a lump in my throat and irrepressible tears blurring my vision. My dusty journey into Idaho's pine-scented past was ended. My comrade naturalists, my kindred spirits, my Long Valley travel companions were gone, like phantoms. In a ghostly shimmer, they receded back into the dim memory of the dusty archives of Baker University. And I was left alone here in modern-day Idaho with no way to look them in the eye, shake their hands and express to them my deep gratitude for including me in their historical summer of adventure. To roughly 99.9 % of Idaho, and equally so to Baker University, they were the forgotten expedition. But I couldn't put Monahan's 1907 trek out of my mind.

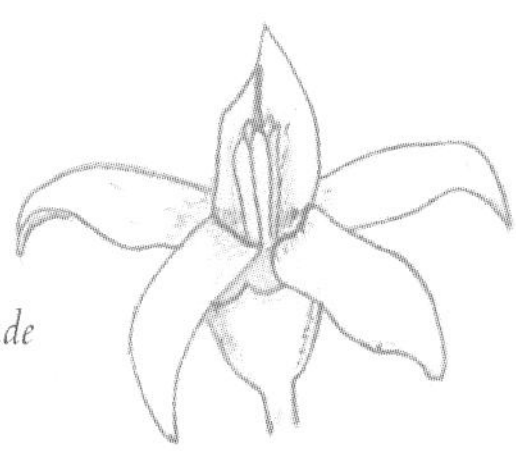

Climbing Nightshade

·2·

PHANTOMS OF THE MUSEUM

From the day of my first reading of Monahan's published diary in 1973, until August of 2009, that diary's narrative was the only known piece of written record of the 1907 expedition's specimen acquisition. Yet, it was blatantly incomplete. The diary records only half of the expedition's journey. It reports very few of the hundreds of bird specimens purportedly collected on the expedition. And even those occasional, scant diary references to particular specimens collected on the expedition were vague or open to interpretation. For example, references to bluebird, hummingbird, merganser, owl, woodpecker, nuthatch, and jays, all beg the question: which species of each? Other references such as red-throated, red-head, and pheasant long baffled me. The "red-throated" bird might have been one of the two woodpecker species with a red throat – the red-naped or Williamson's sapsuckers – both resident along the expedition's route. Was "red-head" (sometimes written red head) a reference to one of the other species of woodpeckers bearing red crowns or faces? Or was it the male western tanager which also bears a reddish or red-orange head? Or were they alluding to the red crown of the Cassin's finch? There is a redhead duck, but most places in which the expedition encounters their "redhead" quarry doesn't square with where you'd expect this duck to be found. And certainly, their reference to climbing a tree to collect a red-head's nest doesn't figure right for a ground-nesting waterfowl. The pheasant collected on July 5 near

Lardo (McCall) – was it a sharp-tailed grouse? But that choice suffered some second thoughts after I learned that sportsmen began introducing the Chinese ring-necked pheasant in parts of Idaho in the late 1880's. Were there introduced pheasants in the farmland around Payette Lake in 1907?

What about the vast number of other specimens *not* mentioned in the diary account? Socolofsky's unconfirmed estimates of the 1907 expedition's acquisition numbers were "250 plants, 1000 insects, 3000 birds and many mammals." Where are those records? Obviously, the vast majority of specimens taken that summer were never noted in the dairy. But that should come as no surprise. After all, diary keeping was intended more for recording general daily events. Collected specimens would have been recorded separately in meticulous, accurate notes, detailing in the very least the species, date, and locality of each collection. Such is the way of science. The burning question remained: where are those collection records?

After reading this diary account, anyone with the slightest inclinations of a naturalist immediately ponders the irrepressible question: exactly what *did* the Baker expedition collect – specifically, in detail – during this biological trek into Idaho? Does a tangible record of their collection still exist? Clearly, the purpose of Monahan's expedition that summer was to gather biological specimens for the Baker University Museum. Did those specimens still reside in their museum? Surely, even if some of the collection no longer existed, there had to have been a catalogued record of the museum's acquisitions. Did that museum record still exist? These questions loomed large and sparked my interest in locating the official written records of the 1907 Baker Expedition.

Those questions were especially tantalizing to me as I felt the need to report this expedition's findings to the scientific community. After all, how many biologists do you suppose come across this account in the 1971 issue of the Idaho Historical Journal? Someone needed to get that scientific information – the bulk of which was about birds – into a broader source of biological readership. I pored over the diary account, attempting to glean from it every discernible bird species they recorded. By so doing, I compiled a list of 27 birds that could be positively identified from the Socolofsky's published record of Monahan's diary. In addition to the many references to birds, the diary also makes occasional mention of mammals, as well as rare references to the collection of insects and plants. Still, the question lingered: *exactly what wildlife specimens* – not mentioned in the diary – did the Baker team take from the 1907 Idaho countryside? Obviously, the answer lay hidden in the Baker University Museum. Only the museum's records could disclose the full story of the 1907 expedition's scientific collection. This museum's catalogue record – residing far away in a Kansas museum – became a scientific sort of Holy Grail, sought after first by Socolofsky, and later by me.

Soon after his discovery of Monahan's diary, Socolofsky visited Baker University to gather background information on both the institution and the expedition's principal characters. Though he was met with the fullest cooperation and assistance, what he found at Baker probably reflected a sad but accurate commentary of the current affairs at many small institutional museums across our country. The popularity of traditional museums was waning. In those late 1960's and early '70's environmental awareness was growing and the old "shotgun school" of collecting specimens was being replaced by more enlightened methods of studying living forms. Field techniques such as live-trapping and tagging of small animals, mist netting and banding of birds, radio-tagging and radio-tracking of large animals were all replacing old-school museum collection. In practical terms, better field guides, improved binoculars and vastly improved photography eliminated much of the need for a dead specimen in the hand to identify critters. Biological field study had advanced from collecting wildlife to connecting with wildlife.

At the time of Baker University's founding in 1858, and for several decades thereafter, museums experienced widespread popularity, especially here in the United States, where there was still so much to be discovered and learned about our country's flora and fauna. As outlined by Scott Weidensaul (2007) the late nineteenth century was agog with a worldwide mania for natural history collections. These holdings were not only among public museums, scientific institutes, and academic institutions such as colleges and universities, but many households sported their own personal collections, sometimes displayed as elaborate and decorative home decor.

Certainly, Baker University stood as an example of the early fascination with scientific collections. A small institution, as Universities go – Baker's annual enrollment from 1905 to 1910 ranged between 341 and 500 – yet, by the early 1900's their museum collection numbered more than 80,000. By 1909 a Baker report touted its holdings as 131, 506 specimens. Among Dr. Socolofsky's notes (generously lent to me) is a reference to the 1945 estimate of Baker's museum holdings at 139,139. In all likelihood, from that time on, the museum lapsed into declining importance on the Baker campus.

At Socolofsky's first contact with Baker in the fall of 1966, there were hints that their new college president seemed interested in reviving the museum and restoring the dusty old specimens to their original condition and location. However, on Socolofsky's first campus visit, when he was taken to the attic-like top floor of one of the buildings to view some of the old museum specimens, he overheard his Baker docents discussing plans to get rid of almost everything due to lack of space, and longtime neglect.

In 1973, after I first became aware of the published account of the 1907 Monahan Expedition, I was immediately interested in tracking down any sort of record of the Idaho trek. The diary's snippets of what they observed and collected along the way only whetted my naturalist's curiosity to know more. I wanted desperately to know *everything* that was collected that summer.

I had shared the diary publication with Dr. Charles Trost, my Idaho State University graduate professor and mentor. He concurred that the biological findings of the expedition needed to be reported more directly to the scientific community in an appropriate scientific journal.

We decided to contact Baker University regarding the matter. Because Dr. Trost wielded more professional clout than I (a mere graduate student at the time) he wrote the Baker University Biology Department in December of 1973, requesting any details they might have regarding the precise specimen acquisition from Idaho on the Monahan 1907 Expedition. Trost received no immediate reply, so a month later I followed up with a similar request.

In my letter I reiterated our interest in reporting the expedition's findings to the scientific community. I alluded to the scientific value of the expedition's historical sampling of Idaho's wildlife, and how Idaho's remote, rugged landscape, had invited precious few pioneer scientific samplings of its wildlife past.

Soon afterward, in the spring of 1974, I received a cordial response from a Dr. V.E. Nelson, Chairman of the Baker University Biology Department. Offering his apologies for his belated reply, Dr. Nelson was apparently unaware of the old expedition and expressed interest in this snippet of Baker University history. He promised to look into the availability of any museum records and specimens from the 1907 expedition.

With this began a long, punctuated correspondence (inherent to our like circumstances of both being professors in a small college where workload and responsibilities regularly exceeded the time and resources available to accomplish our given tasks). For several months I heard nothing further from Dr. Nelson. So, appreciating his probable busyness, on April 22, 1975 I attempted to revive our correspondence by sending him a copy of Socolofsky's report of the Monahan diary. With the article I also inquired about his success in locating any information on the 1907 museum specimens.

I heard back almost immediately, receiving what appeared to be a hastily hand-scribbled letter dated April 30, 1975. In his note he reported that he had received clearance form the Baker Administration to provide me any specimens collected in Idaho by the Monahan Expedition – *on permanent loan,* if we wanted them. Looking back now, the prompt note and its generous offer of the specimen's "permanent loan..." signaled, ominously, the dire status of the Baker Museum. Dr. Nelson's letter seemed hurried and harried, and he signed off with his urging:

> *"Please jog my memory again in June, if you haven't heard from me. There is a lot happening now!"*

Awash in my own frenetic labors of trying to finish my Ph.D. dissertation amid my rookie year of teaching at Northwest Nazarene University, I failed to get back in touch with Dr. Nelson. A full year elapsed without word from Baker University. For the most part, I was too busy to notice. Finally, with my own head above water, on April 30, 1976 I sent a letter to Dr. Nelson offering my confession of neglecting to remind him the previous year about sending the Monahan museum specimens.

Three days later I received the following letter from Dr. Nelson. It intoned the gravity of recent departmental turmoil and the pressure that he was laboring under. And clearly it sounded the death knell of the Baker University Museum. Again, his letter appeared hastily penned in long-hand on plain departmental stationery:

> *5/3/76*
>
> *Leon:*
>
> *We have moved twice since I last wrote you about the birds. They are presently on their way to the auction block, so you wrote just in time. We have no place to house them. We will ship you all of the Idaho birds. (Only 3-5 specimens as I remember)*
>
> *Would you be interested in the rest of the collection of mounted birds? (About 50 from early 1900's and a few from the late 1800's. They will go on the auction block in the latter part of May. Probably won't bring much. Let me know where to ship.*
>
> *Sincerely,*
> *V. E. Nelson*

Despite a glimmer of hope for my long pilgrimage toward obtaining museum information from Monahan's trek, this communication bode dismally dark for the future of the Baker Museum. I was struck with a sense of profound sadness. The museum of the institution's glory days – of Monahan's era of Spartan expeditions and holdings enrichment – was being gutted out of financial expediency. At a time of financial strain, and institutional restructuring, no one could fault the university for phasing out a moribund, long-neglected facility which had outlived its usefulness. I understood Baker University's actions with the museum. Though twelve hundred miles removed, from a site I'd never laid eyes on, I felt the pathos of loss of the museum's legacy. I envisioned the lingering ghosts of Monahan and Will Green presiding sadly over the dismantling process. With each discarded museum specimen a shimmering scene from Idaho's pristine past blinked out. The tossing of each bird specimen extinguished – like a Fourth of July sparkler sputtering dead – a vista of Idaho's history: at "Ruby" Creek, Placerville, Dry Buck Mountain, High Valley and Payette Lake. Among the feathers of the old bird specimens still lingered traces of Idaho past – dust and elements encoded by the Idaho landscape – mineral imprints of long stretches of old wagon trails.

The demise of Baker's Museum was clearly signaled by Nelson's offer of the remaining bird specimens – at no cost – except for the shipping. I was taken-aback by that offer. NNU has never had a museum. Sure, we have had a small "teaching" collection of study skins of various birds and mammals – this, largely through a donation from former mid-1960's biology professor, Dr. Walter R. Quanstrom. Study skins, also known as museum skins, are the stuffed remains of collected specimens, laid flat with legs outstretch if a mammal, or lying on its back with wings folded next to body, if a bird. For ease and to minimize storage space, these skins are not mounted in standing or wing-spread form, as the more life-like, natural-looking taxidermy forms. And within our own cramped

Biology Department, a precarious fate awaited these half-dozen specimen cabinets crammed with study skins – themselves periodically displaced to the next available vacant corner. This collection of study skins could hardly be called a museum. All we had were several display windows along the halls of our biology floor in which a number of donated taxidermy mounts, mostly birds, resided.

The irony of Dr. Nelson's generous offer of free specimens was that we at NNU were experiencing the same space crunch and limited funds as Baker University. As I considered his desperate offer, I felt painfully helpless to accept. Moreover, I was appalled at the scant few remaining 1907 Idaho specimens – "Only 3-5 specimens..."! Only that, from what surely must have been several hundred collected by Monahan on his summer expedition. Obviously, much of the old museum's holdings already had been discarded. Though separated by many miles and institutional tradition, I felt saddened by this lost link between Baker University and Idaho's past.

I responded immediately to Dr. Nelson requesting that he ship to me C.O.D. any Idaho specimens he could rescue from the auction block. I also made one final plea for any museum records or list pertaining to the specimens of the 1907 expedition.

On May 14, 1976 I received a brief, hand-written note from Nelson stating that he would be mailing the surviving 1907 specimens in the next day or two. Two months later, on July 17, 1976, I received the package of specimens from Dr. Nelson. As I gingerly and reverently unboxed the tattered 1907 specimens, with their tags now browned with age, I set them before me on a lab table. One by one I removed the following six specimens and one separate specimen tag:

I "Lousiana" [Western] Tanager (male) *Pyranga ludonieiana* [*Piranga ludoviciana*] – Ada County, June 6, 1907. (Specimen No. 34, on a tag attached to its leg)

I "Canada" [Gray] Jay (partial specimen) *Perisarens Canddensis* [*Perisoreus Canadensis*] – Boise Co., Idaho

August 1, 1907 (printed on a rectangular tag glued to its stand).

1 Willaimson's Sapsucker [*Sphyrapicus thyroideus*] (male) (no other data with it)

1 Lewis's Woodpecker [*Melanerpes* lewis] Ada Co., Idaho, May 29, 1908* (printed on a rectangular tag glued to its stand) (Specimen No. 10, on tag attached to its leg) * Error on year—Field Journal shows two of these collected May 29, 1907!

1 "Arctic Blue" [Steller's] Jay – *Lidia Arctica* [*Cyanocitta stelleri*] Male, Boise Co., Idaho July 11, 1907 (Specimen No. 89, on a tag attached to its leg)

1 Spruce Grouse [*Falcipennis canadensis*] (labeled only, "Female")

1 Rectangular printed tag labeled "Black-headed Grosbeak" [*Pheucticus Mealanocephalus*] Boise Co. Idaho, June 13, 1907

For several minutes I sat, alone in the stillness of the deserted lab room and studied the displayed relics. In a whimsical way, I suspected I was not really alone; that the unseen spirits of Elmer P. Monahan and Will Green hovered over those specimens with me. For if spirits exist, as I am inclined to believe, there were at least three of us honoring those enduring, feathered echoes from the past. As vividly as their feel to my fingers' touch, they ushered me back to another time in Idaho, to the clank and squeal of wagon wheels, and the thrum of insects unchecked by DDT. The "surviving seven" melded my spirit with the wind that stroked the wild landscape of 1907, whisking me with it up the hillsides of Robie Creek, over the purling waters that danced down canyons not yet roiled by mining dredges. From the stolid stare of each of the immobile birds emanated a unique story of a former Idaho landscape upon which they once lived, breathed and sang to the tree tops, sounding their own Idaho anthem. These Monahan relics gave me a musty touchstone to Idaho's summer of 1907. I softly stroked the same feathers that he had so carefully groomed into place at his 1908 museum workplace. Each touch summoned a fleeting sense of the presence of Elmer Perry Monahan and his forgotten expedition.

Though most grateful for Dr. Nelson's enormous efforts in rescuing the seven 1907 expedition remnants and sending them to me, still, no museum catalogue or other official records of its holdings had been found. Such a document became all the more important now, in light of the scanty specimens actually surviving from the 1907 trek. In my desperation for that elusive information I wondered if Socolofsky, in the course of his original research of the Monahan diary, might have obtained any such museum records?

Soon after receiving the "surviving seven" specimens of the 1907 Monahan Expedition, I wrote Dr. Socolofsky at Kansas State University as a last gasp effort to uncover further information on the expedition's acquisition. I hoped that perhaps he had acquired some of the detailed museum data, especially pertaining to the 1907 trek. Though I wasn't even sure the historian still resided at Kansas State, I was pleased to receive an immediate reply from Socolofsky. Regrettably, he had no specific museum data either, but generously enclosed, on temporary loan, his packet of notes taken while visiting the Baker campus and researching background information on Monahan and the 1907 expedition. He also provided me names of a couple of people on the Baker campus who might be able to help me uncover some of the museum information I was seeking.

Following those leads, on August 12, 1976 I made inquiry to Dr. I. L. Boyd, a long-time biology professor at Baker who had been Socolofsky's contact and collaborator during his research of Monahan's diary. I made my usual appeal for any old Biology Department documents that might detail the museum's records. Because it was now a decade since Socolofsky met with him, the uncertainty of

Dr. Boyd's continued tenure at Baker prompted me to send my letter to him in care of the Baker Registrar's Office, assuming they could surely track him down on campus or elsewhere.

At the same time, I wrote the Baker archivist, Mrs. Maxine Kreutizger at the United Methodist Historical Library on the baker campus and asked for any relevant museum information or reports from the 1907 expedition. I hoped that she and Dr. Boyd might collaborate in their search for my requested information. After a month of checking the library's archives and other local sources, the report came back that the campus search produced nothing new. I was profoundly disappointed when Mrs. Kreutziger's letter of September 13, 1976 informed me that my prospects of finding any kind of a historical record of the Baker Museum's holdings were, ... as she put it,... "hopeless." Nor did I hear anything further from Dr. Boyd in the Biology Department. With that, I gave up, feeling I had exhausted all available means of exhuming the ghosts of the 1907 Idaho trek.

After those four years of failure to recover the lost 1907 Expedition collection data, in 1977 Dr. Trost and I published (Powers and Trost 1977) our compiled list of 27 birds that we had gleaned from Monahan's diary. It was a mere sliver of the actual specimen acquisition, but at least the scientific community was made aware of the last of Idaho's "pioneer" biological expeditions.

Uncannily, the 1907 Baker University Expedition story refused to be laid to rest. Many years later, a chance encounter far from Idaho prompted my retelling of its story one more time. It happened in late May of 1991 while I was in Chicago attending a conference titled "The Teaching of Ethics in College Curricula". Seated in the midst of a large crowd of a packed convention center, I was privately resentful of my being at this conference. It had little to do with my teaching background or interests. Heavy on my mind were the gazillion things that needed to be done back home. Admittedly, I was enjoying the company of my colleagues, Drs. Steve Shaw and Dan Berg. But at the moment I was perusing the conference's program of topics and speakers, desperately searching for something that might be of borderline interest to me. Then an item on the agenda caught my eye. One of the featured speakers was from Baker University. Aha! I began plotting a quick antidote for these ho-hum meetings. Since the Baker representatives were reporting on a special capstone ethics class taught by their Science Department, one of the presenters was Professor Don Hatcher from their Biology Department and the other gentleman was Dr. J. Keith Keeling, Provost & Dean of the university's College of Arts and Sciences. Wow, I thought, a modern-day biologist from Baker University. Naturally, I wondered if they knew anything about their old museum, or about the 1907 Monahan Expedition. I was fairly certain they hadn't the foggiest notion of either. Of course, I wondered too, if they would be even vaguely interested in that piece of their history. As they gave their presentation that day, I scrutinized them carefully so that I might later recognize them in the large crowd and perhaps speak with them during a session break.

Later that day during one of the meeting's coffee breaks, that opportunity presented itself. I singled out Dr. Keeling and introduced myself, explaining that I was from Idaho and asked him if he was aware of the Baker University-Idaho connection. I could immediately tell by the way he cocked his head and half-squinted his eyes that he was genuinely curious.

"No…tell me about it", he urged. Briefly, I recounted the story of the 1907 Monahan expedition. I concluded with my lament of being unable to recover any official documents of the old museum's acquisitions, especially relating to that Idaho expedition. Dr. Keeling found this little-known bit of Baker history interesting and asked if I might send him a copy of the published report of the old expedition. In the meantime, he promised to do some checking around the campus for any of the old museum records I sought.

The following week, after my return to Idaho I sent Dr. Keeling a copy of Monahan's published diary. I also included a copy of the bird list that Chuck Trost and I had gleaned from the diary account.

Though not expecting much in return, I was curious to know what kind of sway Dr. Keeling might wield in his search for old museum documents. His interest seemed genuine enough, and with his titles of Provost and Academic Dean at the University, I held out hope that his administrative clout might sweep the archive dust a little cleaner.

Not long thereafter I received a cordial letter from Dr. Keeling. To his credit he had made an earnest effort of sleuthing out some archival campus information on the old museum. It consisted mainly of several 1909 alumni newsletter statements regarding overall numbers of the museum's current holdings. In essence, it was a repeat of the information contained in the notes Socolofsky had shared with me 15 years earlier. And like all former inquiries there were no specimen lists or other details about the 1907 acquisitions. Keeling mentioned that Baker still maintained a small museum, in their oldest building known as "The Castle". It consisted mainly of artifacts from frontier settlement days, including Native American stone artifacts along with materials from frontier farmers and other items of local history. This 1990's scaled-down version of their museum apparently still held a few stuffed animals including birds, probably remnants from the old museum. Regrettably, the lack of identification tags on those few old specimens rendered their origins indeterminable. Once again, I had drawn a complete blank in my quest for the old Baker Museum records. Apparently, even administrative sway held no magic over the dusty catacombs of the old museum.

Buttercup

·3·

FOLLOWING IN THEIR TRACKS

My ever-lingering fascination with the 1907 Baker Expedition, long after first reading about it, prompted periodic scheming of how I might tap again into that experience and revisit their story; perhaps even retrace the expedition's path. But heading out into the Idaho backcountry, following their route posed a formidable task, far beyond my resources of time, energy and finances. Even so, any mention thereafter of that era never failed to catch my attention, summoning me back into the contemporary world and time of Monahan's expedition.

Finally, in 1989, during the ongoing hoopla surrounding Idaho's centennial celebration, an idea struck me. I conceived the notion of teaching a unique summer field biology class. I had taught a summer field biology class once before, but this course I envisioned would be of a much different format. In the spirit of the Idaho Centennial, I would structure the curriculum around a historical theme. This new class would study Idaho's original field biologists – the naturalists conducting those early biological explorations into our state. We would briefly review and map each of those expeditions. But more importantly, we'd also note the fauna and flora that each trek contributed to our knowledge of Idaho's wildlife. That part of our study would acquaint students with the animal and plant life in Idaho's outdoors.

Of course, in this revamped field biology class, the 1907 expedition would be the final pioneer trek studied. Highlighting the

1907 expedition would provide the culminating, hands-on experience for the class. We would take field trips on several days, retracing much of the Monahan expedition. Each day's excursion would follow a different segment of the 1907 trek. As part of our daily studies, in both lab and field work, we'd learn how biologists discover, identify and preserve fauna and flora on such field expeditions. This field biology class melded history and natural history into a relevant whole.

This unique field biology class came to fruition in the summer of 1990, when I taught the first "Early Expeditions" Field Biology Class at NNU. It went so well that I offered the one-week class each successive summer through 1997. I timed the class to correspond to some of the same June and July segments as the original expedition. Class enrollment was always small – six to 12 students – ideal numbers for such a field-oriented class.

As we trekked over sectors of the old 1907 expedition trail each summer we noted our wildlife observations along its path. We then compared our findings with the conspicuously scant data written in the 1907 diary for those same localities. To handle the instructional load and field trip responsibilities for such a jam-packed field and lab-oriented learning experience, I enlisted the able assistance of my two good biologist friends Timothy Craig and Allen Dale. Though they assisted on separate years they each brought to our learning experiences their individually rich and learned backgrounds. Those eight summers of retracing portions of the 1907 Baker Expedition, nearly a century later, provide the underpinnings of this book. At that time, I was still limited to the diary as the lone source of expedition specimens taken on that 1907 trek. That historical diary account laid alongside modern day scrutiny provided an insightful perspective on Long Valley's history and natural history.

The eight summers spent re-tracing the Baker University 1907 Expedition only whetted my intrigue with Monahan's trek. If my 1973 reading of his diary kindled my interest in this phantom trek, retracing its travels through the eyes of a field biologist fanned the flames of fascination. At the beginning and ending of each of those classes, I was struck with the longing to write a book about that forgotten expedition. It seemed such a travesty that Baker University had no lingering knowledge of their long ago connection to Idaho. Moreover, Idahoans knew nothing of that pioneer scientific trek. In a recent inquiry to Arthur Hart, our resident Idaho History guru and writer, he confessed that even he'd never heard of the 1907 Baker University Expedition – despite its 1971 reporting in the Idaho Historical Society Journal.

I longed to put their story into a book, but it would have to wait. Now, more than a decade after I taught the last of those Field Biology classes – three years after retiring from full-time professorship – the waiting was over. Yet, there remained two great challenges in writing this story. First, the expedition leader and central figure, Monahan, was essentially missing in action; I knew little about the man. Secondly, and critically, the scientific bullion of his expedition was seemingly lost to the ages. I could possibly re-discover Monahan, at least learn more about him, but the expedition's mysteriously missing payload was out of the question – or was it?

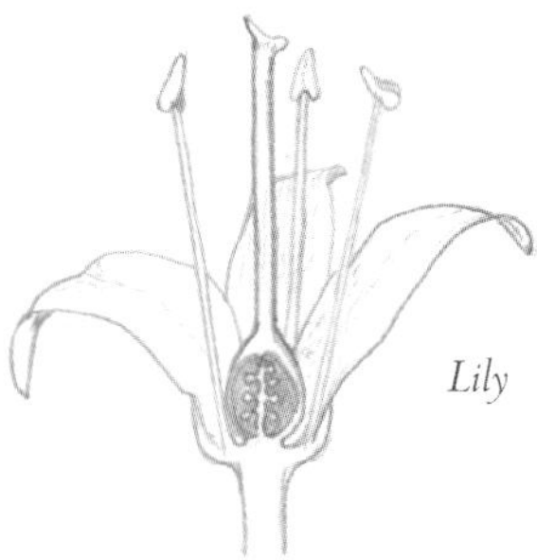

Lily

·4·

FINDING MONAHAN

At the time Elmer Perry Monahan undertook his 1907 expedition he was 33 years old. By 1907 life expectancy standards he had already lived out half of his life. Even so, there still remained many productive years in his life story. In keeping with his low profile beginnings (his 1907 diary didn't even come to light until a half century after the fact), much of his remaining years passed unobtrusively. One has to dig deep to find any record of this man. Socolofsky, (the publisher of Monahan's 1907 diary) apparently found this to be the case. All he provides on the expedition leader's biography is a single sentence at the beginning of the diary article, and a single paragraph at its conclusion. Regrettably, Monahan died just two years before his diary even surfaced, precluding any possibility of a face-to-face interview by Socolofsky.

This cursory treatment of Monahan's biography clunks short and sadly hollow for me. Surely, it ignores the quintessence of the man. There remain so many questions. What about Monahan's life leading up to1907? What sort of childhood did he have? He would have been nearly 30 years old when he matriculated at Baker University. How and why did he end up going to college at a time when pursuing an advanced education was the exception, not the norm. And why attend Baker? Did he have Methodist connections or affiliations? Did his connections with Baker University extend beyond 1910, after he left Baker to attend Medical School? And I can't help but wonder if Monahan maintained his interest in natural history later on, even through his long medical career? And did he have time or interest to continue his taxidermic handiwork? These are but a few questions that hounded me for years.

Finally, in 2009, while writing this book – 36 years after first reading his diary – I reached a point of near desperation to learn more about the man, Monahan. I had never even seen a photo of him. Hence, his name had long flitted through my mind like a faceless phantom. If I couldn't validate his amazing 1907 scientific collection, I would at least find out more about the expedition leader himself.

Where to begin? Where else but on "Google"? I'm not one to commonly surf the internet. After years of poring laboriously over hard-copy, refereed science journals – whose credentialed sources I know and trust – I question the credibility of much of what can be found on the internet. But on this occasion I relented and Googled "Elmer Perry Monahan" just to see what might surface. I wasn't expecting much. But what immediately popped up at the top of my computer screen brought me up short.

The words at the top of the screen boldly read, "Elmer Perry Monahan Jr.", followed by some brief biographical information. This Monahan junior was a retired physician residing in Craig, Colorado. Stretching my computer sleuthing skills to the max, I ferreted out his phone number. All this time the question hung heavily and expectantly: could this be the son of the 1907 Baker University Expedition leader? How many Elmer Perry Monahans could there possibly be, anyway?

Elmer P. Monahan Junior was 89 years old, which certainly fit chronologically. If he were indeed *the* son of *the* Baker University Monahan, he was now, ironically, the age at which the elder Monahan died in 1964. That thought spurred me to further action. Time was of essence.

My next question was equally unsettling: in what physical and mental condition might I find this 89-year-old Monahan Jr.? There was one sure way to find out; I would immediately call this gentleman and resolve the matter. Though reluctant to be too hopeful, I was strangely optimistic that I would soon be speaking with the son of the central figure of the 1907 Baker expedition. I felt a strange new tinge of hope for putting a face on the Monahan of old.

I opted to place my phone call to Dr. Monahan in the late afternoon, after usual nap-times (considering my own proclivity for early afternoon power-naps) but well before the dinner hour. Around 4:30 on the afternoon of January 21, 2009 I called the Monahan number but received no answer. Disappointed but still hopeful, I waited until later that evening and called again. After two or three rings there came an answer.

"Hello", said a scratchy but robust voice.

I hesitated momentarily, questioning just how to begin. Obviously, first I needed to know if I was talking to the right party.

"Dr. Monahan?"

"Yes," came the reply. I was relieved to know I had successfully reached Elmer Perry Monahan Jr. Though impatiently wanting to blurt out the obvious question, I felt I needed to first introduce myself and explain the reason for my inquiry. Doing so briefly, I then asked the question:

"Would your dad have been the Elmer Perry Monahan who led a collecting expedition into Idaho from Baker University in 1907?"

There was a brief silence. I held my breath. It seemed to me that glaciers must have groaned forward several inches in the polar reaches before his answer finally came.

"Yes, possibly; he did occasionally talk about an expedition, maybe two, that he was involved in early in his career. And he had connections with Baker University at that time, too."

I was elated, but also a little puzzled. Apparently this son of Monahan didn't know much about his dad's 1907 exploits, and certainly seemed unaware of his father's published diary. So I put the question to him.

"Are you aware of the publication of your dad's diary? Have you ever seen a copy of it?" I asked.

"No, I don't know anything about that", he replied.

I briefly related its publication date and journal source and then asked him if he'd like for me to send him a copy of the published account.

"Most certainly, I'd appreciate that very much." A hint of mounting interest crept into his voice.

I proceeded to obtain Monahan Junior's mailing address with the promise to send him a copy of the published diary forthwith. Before concluding our conversation I expressed to Monahan how deeply I had enjoyed reading about his father's expedition and related how I had personally retraced much of the trek with my college students. I also asked if he would mind if, along with the copy of the diary, I sent him a few questions about his father. He consented, and we said our cordial goodbyes and hung up.

I sat quietly for several minutes, phone still in my hand, almost in disbelief. It was slowly sinking in – I had just spoken with the living son of *the* Elmer Perry Monahan of the 1907 Baker Expedition! Despite my jubilation, I was surprised that this son had never seen nor had any prior knowledge of the published diary of his dad's historical trek. At most, he knew of the 1907 trip only by hearsay.

Immediately, I set about composing a letter to Monahan Jr., to accompany the copy of his dad's published diary. In the letter I mentioned how I had located him by Googling his name. I also explained how his dad's diary had surfaced inexplicably more than a half century after it ended, and how even then it was only by accident that it landed in the hands of an historian. I also explained how I stumbled onto his dad's diary account some 36 years ago, and how it had piqued my naturalist's curiosity so intensely that it never relinquished its hold on me.

I briefly chronicled my unsuccessful search for his dad's museum data and Idaho specimens from the 1907 trek. I recounted the summer field biology courses that studied and retraced his father's Idaho journey as a means of vicariously reliving that historical summer trek.

My letter ended with a cordial appeal for any old photos of his father, as well as any old correspondence of his dad's that might be helpful for my book about the 1907 trek. I hoped wistfully – and perhaps quixotically – for some old photos or letters by the elder Monahan from the 1907 expedition. Could such old treasures still reside in the Monahan family archives? What a find and prize such artifacts would be!

I wrote my letter on the eve of my phone conversation with Monahan Jr., January 21, 2009, and mailed it the next day. I wondered how all this would be received by the 89-year-old Monahan. How would this unexpected flashback into his father's past settle in, so late in his life? I sincerely hoped, as I stated in my letter to him, that through this reading of his dad's long lost diary, he might find a warm and meaningful reconnection with memories of his father.

Five days later, on the evening of January 26, I received a phone call from Dr. Monahan. It began rather formally – after all, we were essentially still strangers.

"Dr. Powers, this is Dr. Monahan."

I held my breath, nervously anticipating his response.

"I want to thank you for your eloquent letter and the diary material you sent me," he continued.

It wasn't just his words that assured me. I was pleased to note the obvious appreciation so clearly evident in his voice. He went on to affirm how the diary had opened up a whole new chapter of his dad's life that he had been little aware of. Though Monahan Jr. had read only the first several pages of the diary it had already stirred his excitement and he was eager to revel in more of his father's chronicle of his 1907 summer. With equal excitement Monahan told how he was already digging out old family photos and files in a search of any useful material for me and my book.

We chatted for several more minutes. I asked him if he remembered his dad ever keeping taxidermy mounts around the house in later years. The younger Monahan thought he remembered several for a time but that his dad later donated them to the University of Chicago Field Museum of Natural History.

I'd long wondered if the expedition leader ever returned to

Idaho after that original 1907 trek. So I posed the question to his son. As far as Monahan Jr. knew, his dad never again visited Idaho although he often spoke of Will and Ray Green. He also made occasional reference to his 1907 alias, "Cot".

We soon concluded our conversation and promised each other to keep in touch. I felt like I had struck gold in my quest for a fuller understanding of the Monahan of 1907. Too, I still held out hope that just maybe this son of the expedition leader would yet uncover some of the long lost documents of that old expedition. In any case, my intrigue with Elmer Perry Monahan grew twofold – it was now with both father *and* son.

A couple weeks later, on February 11, I received my first letter from E. P. Monahan Jr. It was hand written in bold black ink on personalized stationery. The cursive writing was clear and easy to read. Along with the letter's brief anecdotes of parts of his dad's life story, there was also an enclosed copy of the oldest photo he could find of his father. The picture was taken in 1918 and shows Elmer Perry Monahan Sr. in his Army uniform, while he served in the Army Medical Corps. Additionally, there were copies of two or three other documents and mementos relating to Monahan Sr.

Thus began a regular and delightful correspondence with Elmer P. Monahan Jr. And though the expedition leader's son provided much useful information, several valuable photos and other documents, he could find nothing directly related to his father's old 1907 expedition. Of course, that was my own wild, improbable pipe-dream for which I'd held out hope until the very end. Once more – this time within the bastion of the Monahan family archives – I had failed to find the missing 1907 expedition documents.

Toward midsummer an opportunity arose to visit Monahan Jr. at his home in Colorado. While on a week-long visit with my daughter and her family near Denver I was only four hours from where Monahan lived in Craig. Anticipating this opportunity I had called Monahan the previous week and asked if he would be in town the following week and if so, would he mind my visiting him.

He enthusiastically agreed to the plan and on June 27, 2009 I drove from Denver to Craig.

The route northwestward from Denver took me through spectacular vistas in the heart of the Colorado Rockies. Quaint and touristy little towns like Idaho Springs (which some history pundits claim may be the origin of my home state's name), Steamboat Springs, and Hayden (a name that evokes another, even older Idaho scientific expedition) were a few alluring sites along the way. I resisted the temptation to stop and explore, instead, I pressed on to Craig after nearly four hours of driving.

Dr. Monahan received me most warmly and graciously. I was introduced to his longtime friend Mary, and since it was lunch time, we drove to the nearby Craig Golf Course for lunch at its locally popular Tin Cup Grill. From our window table we looked out upon the lush expanse of groomed greens bordered by stately cottonwoods and riparian tangles of the adjacent Yampa River. I was struck with the appropriateness of this rustic setting for reminiscing about the 1907 expedition leader.

Following our pleasant and chatty lunch we retired back to Monahan's home.

At this time, I unveiled a special surprise that I'd brought all the way from Idaho to show him. It was one of the surviving six 1907 bird specimens – a Lewis's Woodpecker – which his dad had collected and mounted for the Baker Museum. It was one of the surviving six 1907 specimens salvaged from the 1976 dismantling of the Baker University Museum. Of special interest to Monahan Jr. was the old yellowed specimen tag that hung from the woodpecker's leg. On that tag was the date, Idaho location, and the "E.P. M" initials, penciled in his father's hand writing.

As Monahan junior stood reverently studying the 102 year-old tangible connection to his father, he was quiet. I thought I detected a moistening in his eyes.

We stepped outside into the front yard for photographs. As I snapped pictures of the elder Monahan's two pieces of handiwork

– his son and his century-old taxidermy mount – I sensed a fleeting anachronistic flashback. Before me stood a flesh and blood connection to the Baker Expedition of 1907; and cradled in his hands was a tangible survivor of that old trek. All that was missing were two figures in rustic apparel, with rifles in the crook of their arms. Though unseen, my thoughts ushered "Cot" and "Arthur" into our presence. With them wafted the faint scent of dusty trails and sweaty horses joined with the sweet fragrance of flower-spangled meadows. Did my age-dulled ears really detect bird song in hushed harmony with sibilant sighs of monarch pines? We stood briefly reconnected; a storyline running from 1907 to 2009 – long lost, but now ready for the telling.

After our photo shoot Dr. Monahan took me back inside his home and into his downstairs den. Rolled-up maps hung in racks on the walls, and several others sprawled out over a work table, in testimony to his still active hobby of cartography. In an adjoining room was an ample reading library. There he showed me his father's old copies of Asa Gray's 1889 *Manual of the Plants of North America,* and an 1835 illustrated book on hummingbirds. Monahan Jr. recalled a couple of other bird books that his dad long cherished. One was a beautiful, big copy of the 1937 *Audubon's Birds of America.* The other was a 1937 two-volume set by National Geographic Society, *The Book of Birds.* This latter set may have borne special fondness for Monahan because its co-author was his friend, Alexander Wetmore.

Declining Dr. Monahan's gracious offer to stay the night, I concluded our visit with coffee and a sampling of Mary's special cheesecake. If the delightful prospect of visiting again with Monahan doesn't bring me back to Craig, the allure of Mary's cheesecake certainly could.

Since that June visit at Dr. Monahan's Colorado home, he and I continued to correspond throughout the latter stages of writing this book. He has been both a charming contact and a rich source of information on his father's life story. I feel most fortunate to have found Elmer Perry Monahan Jr., the next best thing to talking to the 1907 Baker Expedition leader himself.

·5·

A BOMBSHELL FROM DUSTY BOXES!

Throughout my three-decade-long frustration of not finding museum records or many 1907 specimens, I occasionally thought about those old Baker Museum specimens which had *been sold* at the 1976 auction. How many of the 1907 Monahan relics were let go – purchased or given away? And to whom? Was it mainly friends of Baker University who rallied around this fund raiser to aid the space- and cash-strapped college? If so, I could never track them down. But it also occurred to me that some specimens may have been snatched up by other Kansas natural history museums. It is not known how far and wide Baker University advertised their old museum auction, but I couldn't ignore the possibility of some auctioned specimens ending up at other institutions.

Well into writing this book, I finally got around to querying that last potential repository of missing Baker Museum specimens – other Kansas museums. On February 9, 2009 I sent an email to each of the remaining Kansas museums: the Schmidt Museum of Natural History (Emporium State University), Sternberg Museum of Natural History (Fort Hays State University), and the University of Kansas Natural History Museum. Because Monahan Jr. had mentioned to me that his dad had at one time contributed several bird specimens to the Chicago Field Museum of Natural History, I queried there too. Within the next two days I received negative responses back from all of the institutions except the University of Kansas, from which I heard nothing. The negative response I received from The Chicago Field Museum was both surprising and a little puzzling, since Monahan Jr. told me he had once visited the museum and therein viewed a couple of his dad's specimens. The Chicago Field Museum's response was a little suspicious, since they stated that they did have specimens from a "C.R. Monahan" – two Lincoln Sparrows collected in Baldwin, Kansas. Baldwin, being the home of Baker University made these records even more suspicious.

Curious about this Chicago Field Museum Monahan record, I emailed back for further details on the Baldwin, Monahan specimens – specifically, on the date they were collected. Interestingly, they were collected 27 April 1908. This would have been soon after Elmer P. Monahan's return from his Idaho-Oregon trek. It matters little for the focus of this book, but I'd wager a16-oz. mocha that the "C. R." may have been mistaken for a vintage Elmer P. Monahan "E. P." scribble on the specimen tag. Or, as my son Ryan suggested, maybe those letters were a smudged "Cot" – Monahan's life-long moniker.

As to any auctioned Baker Museum specimens, I suspect that scattered around the Baldwin City, Kansas area and perhaps beyond, among some of the alumni faithful, there might still reside a few of the old 1907 specimens acquired in the 1976 campus auction. By now, over a century old, unless they have been well cared for, any surviving specimens undoubtedly stand precariously fragile in their silent tribute to Elmer P. Monahan and his phantom expedition of 1907.

All that aside, there still remained the most glaring vexation of all: the lost *specimen records* of the old Baker Museum. For over three decades I'd pondered the whereabouts of that lost treatise of Idaho wildlife gathered by Monahan in 1907. Every time I retraced the same dusty roads he had trekked, the specter of his lost collection haunted me. Innumerable wildlife questions hung over me like the festering clouds of mosquitoes of his day. And now, 102 years later, those questions seemed all the more impossible to answer. But I would stubbornly write the 1907 story anyway, minus its lost mother-lode of specimen data.

Then unexpectedly – incredulously – while nearing the completion of this book's writing, I experienced the most remarkable journalistic windfall. And though this munificent bombshell effectively rectified my 36 years of frustration and failure, it turned on the thinnest wisp of luck – perilously close to a near miss. Two small, incidental afterthoughts reversed the course of my perpetually fruitless search for the 1907 specimen data.

The first of these followed my curiosity about one of Socolofsky's footnotes on Monahan's July 5 bald eagle collection at Payette Lake. He therein states that while researching the diary he saw that eagle specimen still displayed on the Baker University campus. I had noticed that footnote several times before but had never given it much thought since my persistent inquiries about Monahan specimens had never elicited any mention of it. But on August 22, 2009 I decided on a whim to contact the Baker Biology Department one more time and make a fresh inquiry, this time specifically about the old bald eagle mount.

It had been over three decades since my last communication with the Baker University Biology Department personnel, so I went to their website and located the name and contact information of their resident ornithologist, Dr. Roger Boyd. In an e-mail to him, I explained my book project on the old Monahan expedition, and asked if the 1907 bald eagle taxidermy mount still stood anywhere on the Baker campus. Honestly, I doubted that it still existed.

Dr. Boyd responded almost immediately. He gave me a brief but fascinating summary of his own long history at Baker University. His father had also taught at Baker from 1941 until his death in 1982. Then it dawned on me; it was his father, I. L. Boyd, who Socolofsky had mentioned and to whom I had sent a follow-up letter back in 1976, as my last-ditch effort to find Monahan specimens. Given the upheaval of change and reconstruction that the Biology Department was then experiencing, it is almost certain that my letter never reached the elder Dr. Boyd.

Roger Boyd went on to express his uncertainty about the status of the old bald eagle mount. But his next casual comment, sort of an afterthought, nearly knocked me out of my chair! I could hardly believe my eyes as I re-read his brief comment:

> *"I do have a field journal apparently from that expedition or another one around that time. It has a newspaper article taped to the inside cover about them leaving for the expedition. It mostly records what they collected, usually dates, and occasionally locations."*

Wow!!! Could that be Monahan's 1907 field journal – exhumed after all these years? What else could it be, especially if it's from "around that time?" I read Dr. Boyd's message again and again measuring each word for the certainty of hope, while already mentally composing a reply e-mail to ask for more details, or even for the journal itself.

I read on. As the field journal statement had stunned me into full-blown hope, his next statement mystified me. He wrote: "I believe I sent you a description of this back in Feb or March when you had inquired about some of the expedition specimens at the University of Kansas."

Bingo! I suddenly remembered that the University of Kansas was the only museum I had not heard back from regarding my inquiries about old Baker University specimens. To my chagrin I realized I had forgotten to check back on that. But, I also was certain I'd never received any such response. If indeed, Dr. Boyd had replied from the Kansas Museum, how could I have missed his email?

I quickly double-checked past e-mails for any from Kansas University or Dr. Boyd, and found nothing. I then fired an email back to Dr. Boyd expressing my supreme interest in the old journal. I also reported my failure to find a copy of his email response from the Kansas Museum and asked if he might forward a copy if it were still in his "Sent" file.

My mind was racing almost deliriously. Had my 36-year-old quest for Monahan's 1907 collection records finally ended – triumphantly? Did this old journal in question really contain the lost records of each bird, mammal, plant and insect specimen collected in wild Idaho 102 years ago? I felt almost afraid to let myself believe that the old journal Dr. Boyd spoke of could *really* be Monahan's.

Dr. Boyd's next email, later that day, provided further details of the journal that sealed its authenticity. It was undoubtedly Monahan's 1907 field journal! Several days later I received a package from Dr. Boyd. As it turned out there were *two* journals. The more-lengthy one contained all of the bird collection data, 153 notations in all. The last 23 bird collections, beginning on November 20, 1907and running through February 12, 1908, were from Josephine County, Oregon, where Monahan went after departing Idaho. The second old journal recorded the 36 mammals collected on the expedition. As with the bird journal, the last nine mammals also were collected in Josephine County, Oregon, between November 20, 1907 and January 4, 1908.

Stuck inside one of the journals, were several loose pages. One page, tattered around the edges was titled "Birds identified in Idaho during summer of 1907". At the top left-hand margin it was dated, "May 25 to June…" The page listed 26 birds by common and scientific name, with a right-hand column noting each bird's "occurrence" (abundant, common, rare, apparently in migration). This list started on the first day of the Baker Team's encampment on Robie Creek. Although there is no concluding date to that list, I'm guessing it ended with their departure from Robie Creek north to Payette Lake on June 22.

The other five loose pages were bound together at the upper right-hand corner by course black thread – perhaps a 1907 version of a "staple". These five pages contained recorded notes on Monahan's oological (bird egg) collection. The first page actually began in 1906 with three collections from the Baldwin, Kansas area (eggs from the nests of a crow, phoebe, and a black-capped chickadee). The next four pages record 18 egg collections from Idaho on the 1907 summer expedition.

Curiously, at the very back of the mammal field journal is a long numbered list of bird names. There is no title to this list nor are there dates or localities included with the names. I'm of the opinion that it is a cumulative list of the birds that Monahan had prepared into museum specimens.

Those old journals constitute the long missing specimen records detailing exactly what birds, bird eggs and mammals were collected by Monahan and Green on the 160 days of their 1907 expedition in Idaho. They contain no records of any other specimens such as insects, fish, amphibians, reptiles or plants. Dr. Roger Boyd recalls some old boxes of unmounted plant specimens that were discarded long ago. Some of them may have been from the 1907 collection. Even though today Baker University maintains a robust herbarium of plant collections, it contains no known remnants from the 1907 Monahan Idaho expedition.

Monahan's long sought-after field journals, with their wealth of priceless scientific information, posed two immediate problems for me. First, much like Socolofsky described Monahan's diary, the journals are written with an indelible pencil, in hastily scrawled, cursive script. Secondly, as if the writing itself wasn't difficult enough to decipher, many of the bird names, both common and scientific, are archaic and quite different from today's usage. I soon picked up on the general pattern of Monahan's writing in his general script but I needed help on the names, especially the technical or scientific names. What bird names were in vogue in 1907? Where could I look for this information? Like the diary, the journals offer no clues to the scientific reference books Monahan must have used – well, almost no clues.

There is one entry in his bird journal that betrayed two of his bird sources; but this wasn't divulged until July16 of that summer, when Monahan, while recording his notes on a Hairy Woodpecker specimen, makes comparative references to "Coues" and "Reed". Those are huge clues, pointing up two of the early 1900's bird guides: Elliott Coues' 1903 *Key to North American Birds Vol.1 & 2* (Dana Estes and Company, Boston), and Chester Reed's 1906 *Bird*

Guide: Land Birds East of the Rockies and Water Birds, Game Birds, and Birds of Prey (Doubleday, Doran & Company, New York). Whatever else he may have carried with him that summer, unquestionably he had copies of these two sources to help him with his bird identification. For my detective work on Monahan's old field journals, I borrowed both volumes of each source – four volumes in all. Those relic old bird books became my faithful partners in making sense of the recorded usage of archaic bird names, both common and scientific.

Monahan's field journal of mammal specimens presented yet a different challenge. It's quite clear that he was not nearly as familiar with mammals as he was with birds. Never does Monahan resort to using scientific names of the mammals, only common names. And many of those common names used are too general even for species identification. For example, four specimens are identified only as "mouse". There's a good chance they were the ubiquitous North American deer mouse (*Peromyscus maniculatus*), but some may have been one of the equally common voles (*Microtus spp.*). Neither did "Little Chipmunk" narrow down which of the 2-3 species of *Tamius* may have been collected. In fact, one has to surmise (as I did, anyway) that the "Large Chipmunk" is actually referring to the golden-mantled ground squirrel. This squirrel-sized sciurid looks remarkably like a chipmunk on steroids – hence its frequent misidentification as a chipmunk or, due to its large size, "Bullmunk".

Despite the sprinkling of these prosaic mammal names, some can be identified with reasonable accuracy: porcupine, flying squirrel, and jumping mouse, for examples. In any case, his Idaho mammal collection includes only 26 specimens; roughly a hundred fewer than birds.

To my disappointment neither of the two field journals records anything of fish, amphibians, reptiles, insects or plants that may have been taken as specimens. My expectations here may have been fueled by a post-expedition 1908 report in the Baker University 50th Year Catalogue which lists among the museum's specimens 12 reptilian specimens from Idaho and 250 botanical specimens and 500 insects form the Northwestern U.S.

Nonetheless, this extraordinary discovery of Monahan's two 1907 field journals sheds unprecedented light on his historical trek. The journals' seismic disclosure combined with findings from my 36 years of historical sleuthing finally makes possible this most revealing expose` of Monahan's 1907 Baker University Biological Expedition into Idaho.

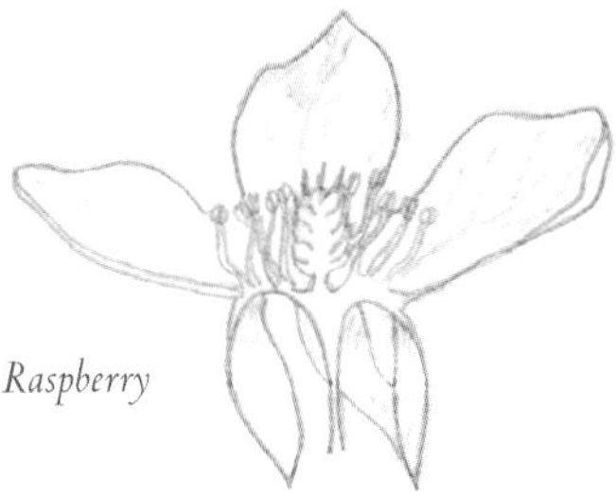

Wild Raspberry

PART II

THE 1907 EXPEDITION

Fig. 206. — Crimson-headed Tanager.

·6·

FROM KANSAS CORNFIELDS TO IDAHO SAGEBRUSH

(MAY 12, 1907)

Elmer P. Monahan and Will Green left Kansas bound for Boise, Idaho on May 12, 1907. The two biologists – naturalists, in the jargon of the day – were on a mission from Baker University, their small Methodist alma mater located in Baldwin, Kansas. From Boise, the two men planned to embark on a summer-long biological expedition that would take them up into the rugged Sawtooth Mountains. Monahan was 33 years old and a 1905 graduate of Baker. Since his graduation he had served on the faculty as an instructor in biology and taxidermy. Plying his skills in taxidermy, Monahan devoted much of his time to preparing exhibits for the Baker museum. This 1907 Idaho trip was the first of a planned series of biological expeditions sponsored by Baker University Biology Department to procure additional collections for their museum. Assisting Monahan on this first excursion was William David Green, a 22 year-old newly minted graduate of Baker's 1907 class.

Though no details are given regarding the duration or mode of travel from Kansas to Boise, other than "they went directly", presumably the team traveled by rail and arrived in Idaho around mid-May. Upon their arrival in Boise they enjoyed a week as welcomed guests in the home of the O.F. Bacon family. That hospitality was no doubt fostered by the friendship of their son, Thomas Clyde Bacon, with Monahan and Green. The younger Bacon was then a student at Baker University, and planned to join the Baker expedition team for several days later that summer. One can't help but wonder how influential this Boise family connection may have been in luring the Baker Museum team all the way from Kansas to Idaho for the 1907 collecting excursion. Why center their expedition in Idaho, and not Oregon, Washington, California, Arizona or elsewhere? Perhaps the Bacon family connection swayed the Baker Team's choice of an Idaho destination.

We do know that Monahan and Green's presence in Boise did not go unnoticed by the local news media. On May 25, 1907, the day they departed Boise for Ruby (Robie) Creek, the *Idaho Statesman* ran a rather lengthy article about the Baker Expedition, outlining its purpose, intended itinerary and proposed mode of operation. In mild journalistic hyperbole it was reported that Monahan and Green would:

"...leave no section of the state unexplored. They will explore the valleys and rivers, will ride the sagebrush plains and travel over the mountains. No part of the state will be too wild or remote... for them to visit. Weeks and perhaps months will be spent in the remotest recesses of the mountains..."

I suspect the proposed acquisitions likewise suffered somewhat from journalistic flair: "...and an effort will be made to secure every living creature. Nothing will be overlooked."

Perhaps Monahan himself invited this overstatement of 'all creatures' with his quoted remark: "...as the mouse is just as important as the moose."

In reality, "every creature" would include a great diversity of animals together with frogs, toads, salamanders, snakes, lizards, spiders. However, there is little indication that any were collected as specimens. Granting a little poetic license, I suppose we can also ignore the newspaper's scientific inexactitude: "to secure specimens of animals, birds, plants and insects." As if zoologically, birds and insects do not qualify as animals.

Hyperbole aside, the newspaper report reveals several interesting aspects of the expedition not otherwise known. It states, for example, the division of responsibility to be assumed by the two expedition members: Monahan would devote most of his time to gathering specimens of "animals and birds" while Green would concentrate more on plants and insects. My naturalist's mind had to again ignore the loose classification of animals; birds are animals, as are Green's sought-after insects. The newspaper report further states that for each specimen collected, Monahan would record the location, altitude, latitude, and general surroundings – habitat, in today's jargon. The expedition also planned to photograph many of the specimen collection sites. None of these details are reported within the published diary account.

With this newspaper sendoff, the Baker Team took leave of the Bacon Family hospitality and headed for the hills where they set up camp 12 miles north of Boise in Ada County on what was then called Ruby Creek, now known as Robie Creek. There they settled for nearly a month, from May 25 to June 22. Absolutely nothing is recorded in Monahan's diary during this hiatus on Robie Creek – until the day they leave for the high country. And even then, little is divulged. We do know that during their Robie Creek encampment, Monahan acquired a light wagon and a pair of old horses from an area farmer. The two horses were apparently well past their prime; one being described as an old sorrel with one eye and one game hind leg, and the other as a bay, "fossilistic in age". By the time the expedition pulled out of Robie Creek and headed north to Payette Lakes, Monahan had also acquired a six-month old dog, a pure white English Pointer that went by various names depending upon its behavioral compliance, but usually answered to "Jo" or "Bughead".

Very likely, part of the stay on Robie Creek was spent readying the wagon for the special expedition. At least a few minor alterations would be necessary for fitting the wagon to the task of hauling collecting gear and storing specimens accrued along the way. In addition to their declared cargo of regular camping gear, a bucket, shovel, two heavy sacks "full of stuff", and two egg crates full of provisions and books, they must have had numerous containers for small specimens as well as hand nets for collecting insects. In the diary entries there are occasional references to a plant press for taking and preserving botany specimens. For acquiring some of the larger specimens they sought, in addition to traps, the men were armed with a .22 Marlin, a .40-82 Winchester, and a .38 Iver Johnson. Understand, those were the days when wildlife study was dominated by the shotgun school, with its accompanying jingle: "what's hit's history; what's missed's mystery!" If a biologist wanted a really instructional look at an unknown critter, about the only feasible way was to "collect" it, which usually meant, shoot it. Telephoto cameras, live-trapping techniques, fiber-optics and other gadgetry were not yet well perfected as research tools. Collected specimens still ruled natural history study.

Photography was to play a part in documenting certain aspects of the expedition. The newspaper article mentions this and at a couple points in mid summer, the diary records Monahan's tinkering with a camera. On June 10, while still at Robie Creek, Monahan describes in his field journal how he photographed a hummingbird in its nest. But little else is reported about photography efforts, and I know of no evidence that any photographic records survived this expedition.

As I ponder the scientific gear and other accouterments available to the Baker team in 1907 for their collecting and identifying of animal and plant specimens, a question looms large: what about books? Among the egg crate of books they claimed to have had, were there reference books for identifying birds, mammals, reptiles, amphibians, insects and plants? Nowhere in the diary account is there any mention of literature resources that Monahan and Green may have consulted in their identification of wildlife they encountered. The two biologists were a long way from Kansas, their homeland and center of academic training. Surely, the wilds of Idaho produced numerous life forms unfamiliar to them. As I trek along with them through their diary account, I read many references to their collection of not only birds and mammals, but insects and botanical specimens as well. Most of the birds mentioned in the diary are identified by common name. Only occasionally is there a reference to observing or collecting a "strange one", presumably new and unfamiliar to them. Similarly, such unknowns encountered on the Lewis & Clark Expedition were usually dubbed "curiosities".

In 1907, what kinds of reference books, bird books for example, would have been available to Monahan? More to the point, were there such things as field guides like we find in abundance today? I suspect that most folks today would assume that the handy bird guides we now enjoy didn't exist until 1934, when the Father of North American field guides, Roger Tory Peterson, produced his first edition of *A Field Guide to the Birds.* Yet, as well chronicled by Scott Weidensaul (2007), the turn of the century saw the appearance of several prototypical "field guides" to birds. The most influential is thought to have been Elliott Coues's *Key to North American Birds,* first published in 1872 and surviving five editions by 1903. Soon thereafter, one of Coues's collaborators, who also was one of the first influential women ornithologists, Florence Merriam, published *Birds of the Village and Field* in 1898 and *Birds Through an Opera-Glass* a year later. One of the next important bird books was Frank Chapman's 1903 *Color Key to North American Birds.* Lastly, in 1906 Chester Reed produced a pair of small but very popular bird books, each simply entitled, *Bird Guides.* The two books contained color paintings and each featured different bird groups.

Another young naturalist, Roy Chapman Andrews, (a contemporary of Monahan's) would later become the iconic explorer and expedition leader for the American Museum of Natural History. In his formative years, Andrews relied heavily on his tattered copy of Frank M. Chapman's *Handbook of Birds of Eastern North America* (Gallenkamp 2001). Might this have been one of Monahan's bird identification primers as well?

This turn-of-the-century proliferation of bird books suggests some possible reference sources available to the 1907 Baker team.

Which of these may have been used by Monahan and Green in their academic training or on the expedition is nowhere stated. It isn't until later in this story that a couple of cryptic clues emerge from the yellowed pages of Monahan's old field journals.

There is a far more tangible clue to Monahan's possible reference book for botanical specimens. This revelation came from a more direct source than inferences from Monahan's 1907 jottings. And it came to light a century later during my recent and enthusiastic acquaintanceship with Monahan's 90 year-young son, Elmer Perry Monahan Jr. In 1931, when E.P. Monahan Jr. was just 12, his father bequeathed to him a long-treasured copy of the 1889 *Gray's Manual of Botany*. His father's love for the book is clearly evident in the inscription he wrote to his son on the book's flyleaf:

"A little message to my son — E.P. Junior-

Dear boy: Between the covers of this volume hides a story of wondrous beauty.

It tells about hills and ravines — all about the things that grow there and why they choose these places in which to dwell — a brook and a meadow — buttercups and cowslips — rock pools and mossy ledges — sprawling vines, ferns and lovely fragrance------

Beautiful and orderly is the account of family, genera and species. Sometimes only a single flower makes up the group but just over the hill dwells a household of many types.

I want you to keep this old book & love it as I have loved it. It will help you understand and appreciate more the rich out doors."

EPM

Interestingly, Elmer P. Monahan Jr. did not discover the inscription on the book's flyleaf until many years later. Today, in the son's words: "I value it more than anything else I have. It defines his life." Also found among his dad's old books, the younger Monahan still possesses an 1835 book on hummingbirds.

Apparently, these are the two remaining relics of the elder Monahan's natural history reference library. The old botany book plus the diary's occasional reference to plant presses supports the notion that the Baker Expedition surely collected floral specimens as well as fauna.

No doubt part of the Baker team's lingering on Robie Creek was to allow sufficient time for early summer snow-melt, and the drying out of roads, and improved travel conditions in the high country. But while this month-long encampment in the Boise foothills gave them time to equip themselves, and assured them better roads; it also quickly endeared them to the Idaho landscape. Their fondness for the idyllic setting of their Robie Creek encampment is evident in Monahan's opening paragraph of his diary:

"...a beautiful little stream clear as glass and cold as ice. This was such a splendid little stream, that we became attached to it and were sorry to have to leave its dark sparkling waters; then it was our first camping place you see and it was like leaving home."

Although the diary says not a word of wildlife encounters while encamped on Robie Creek, Monahan's recently discovered field journals reveal that the expedition immediately launched into its business of specimen-collecting. Between May 25 and June 14, the collection of nearly 50 birds, 14 sets of eggs and nine mammals were recorded in the field journals from that location. Monahan's loose-page bird list titled: "Birds identified in Idaho during the summer of 1907", bears the inscription near the top in the left hand margin: "Ada Co. May 25 to June..." Presumably this list of 26 bird species records what they encountered while camped on Robie Creek. How does this bird list compare with the field journal's numbered entries of collected birds? To answer this question we must take our first peek into the elusive, recently found field journals.

Beginning on May 25, 1907 (their arrival at Robie Creek) Monahan's field journals of bird specimens, egg collections and mammal specimens read as follows (I've tried to present Monahan's journal notes in the manner they are recorded; present day taxonomy and nomenclature are in brackets.)

~~~~~~~~~~~~~~~~~~~~~~

(Birds) No. 1 5/25 Asyndesmus torquatus [Melenerpes lewis] (Lewis['s] Woodpecker)
l of body 3.5; cir of body 5.75 Eyes black
Taken on Mt. side from pine Eggs partly formed
Ada Co Idaho – RubyCreek

No. 2 5/25 Papillo [Pipilo]maculatus articus (Arctic Tohee) [Spotted Towhee]
l of body 2 in very common
Ada Co Idaho Ruby Creek

Also on May 25, 1907 the first mammal specimen is recorded; presumably a Columbian ground squirrel [*Spermophilus columbianus*]:

No. 1 5/25 Mountain Squirrel

The bird collection notations continue in order of dates:

No. 3 5/26 Dendragapus obscurus (Dusky Grouse)
l of body 6.75 – Cir 10.75 – Neck 4.5 – Eyes orange eggs about half formed Flushed her on Mt. side and she flew down into clump of pines where she sat until we shot her. Saw 3 today – all solitary. Found shell of egg of this bird but no nest.
Ada Co. Idaho Ruby Creek.

No. 4 5/27 Stellula calliope (Calliope Humming Bird) [Hummingbird]
Saw 4 took two with dust shot – Not at all shy. All seen on the Mt. side.
~~~~~~~~~~~~~~~~~~~~~~

Ada Co Idaho Ruby Creek

No. 5 5/27 Stellula calliope Calliope Humming Bird [Hummingbird]

Notes as in # ____

No. 6 5/27 Cyanospiza [Passerina] amoena Lazuli Bunting

l of b 1.5 Saw pr. on Mt. side took only

Ada Co Idaho Ruby Creek

No. 7 5/27 Falco sparverius (Sparrow Hawk) [American Kestrel]

Taken from top of tall pine on Mt. side

Ada Co Idaho Ruby Creek

The first two sets of eggs are collected on this date as recorded in Monahan's oology notes:

No. 1 5/27 Spizella socialis [passerina] (Chipping Sparrow)

No. in set 4 Pos[ition] – *Pine tree 5 ft. from ground on Mt. side.*

No. 2 5/27 Papillo [Pipilo] maculatus arcticus (Arctic Tohee) [Spotted Towhee]

No. in set 4 One egg hatched & others ready to come out. Got parts of two only. Pos[sition], *on ground in clump of sagebrush.*

May 27 is a rather productive day for Monahan. In addition to the above four birds and two sets of eggs, he adds two more mammals to his collection:

(Mammals) *No. 2 5/27 Pine* [Red] *Squirrel [Tamiasciurus hudsonicus]*

Taken from nest in willows along creek bank 10 ft. high: nest

of vegetable fiber 8 x 10. This type is quite common here
Ruby Creek Ada Co. Idaho

No. 3 5/27 Pine [Red] Squirrel [Tamiasciurus hudsonicus]
Notes as in # 2

(Eggs) *No. 3 5/28 Dendragapus obscurus (Dusky Grouse)*
Found part of shell on Mt. side

(Birds) *No. 8 5/29 Falco sparverius (Sparrow Hawk)* [American Kestrel]
Ada Co Idaho Ruby Creek
No. 9 5/29 Asyndesmus torquatus [Melanerpes lewis] (Lewis['s] Woodpecker
Ada Co Idaho Ruby Creek

No. 10 5/29 Asyndesmus torquatus [Melanerpes lewis] (Lewis['s] Woodpecker
Notes as in # 9

(Mammals) *No. 4 5/29 (Mountain* [Northern] *Flying Squirrel) [Glaucomys sabrinus]*

(Birds) *No. 11 5/30 Merula migratoria (~~Robin~~ propinqua [Turdus migratorius]* [American Robin]
Taken on Mt side
Ada Co Idaho Ruby Creek

No. 12 5/30 Dendragapus obscurus (Dusky Grouse)
Flushed her from the Mt. side where she had nest under clump of Sage Brush ([left blank for later entry of scientific name] *)*
No of eggs 7 all fresh. She flew to nearby tree where we shot her with 22 rifle.

Body measurements same as in # 3
Ada Co Idaho Ruby Creek

No. 13 5/30 Pyranga [Piranga] ludoviciana (Louisiana [Western] *Tanager)*
Taken from a pine tree ([left blank for later entry of scientific name]*) Song much like Baltimore Oriole (Icterus galbula)*
Ada Co Idaho Ruby Creek (Mt. side) l of b 2 in

No. 14 5/30 [blank – possibly a second Western Tanger?]
Taken on Ruby Creek Ada Co Idaho

No. 15 5/30 Zamelodia melanocephala [Pheucticus melanocephalus] Black-headed Grosbeak
Taken in tall pine tree. Song beautiful, resembling song of Habia ludovicianus (Rose-breasted Grosbeak). The only locality in which this bird was seen. Two others were heard in same locality.
Ruby Creek Mt. Ada Co. Id.

No. 16 5/30 [blank...Western Wood Pewee?]
Taken in a ravine, song resembling the Contopus virens ([Eastern] *Wood Pewee) Very common.*
Ruby Creek Ada Co Id.

No. 17 5/30 [blank]
Notes as in # 16

No. 18 5/30 Stellula calliope (Calliope Hummingbird)
Ruby Creek Ada Co. Id.

(Eggs) *No. 4 5/30 Dendragapus obscurus (Dusky Grouse)*
Taken on Mt. side alt[itude] *800 ft. No. in set 7 – all fresh*

Measurements 1.90 – 2.00 – 1.80 – 1.75 – 2.10 – 1.70 – 1.95
Two half formed eggs taken from bird. Nest a few leaves and pine needles under a Sage Brush (blank *). Behavior of bird- not shy; when perched in tree seemed to feel safe and at ease. When flushed made a terrific noise - so sudden that I could not or did not shoot at her on the wing. Green shot her with 22 rifle.*
Ada Co. Id Ruby Creek Mt.

No. 5 5/30 Pipillo maculatta [Pipilo maculatus] arcticus (Arctic Tohee) [Spotted Towhee] *Number in set 5*
Taken on Mt. side; nest on ground
In clump of Sage Brush ([blank] *) Not incubated*
Ruby Creek Mt. Ada Co. Id.

(Birds) *No. 19 5/31 Dendragapus obscurus (Dusky Grouse)*
Taken on Mt. side. Ruby Creek Ada Co. Idaho
Size 19 inches l

No. 20 Stellula calliope (Calliope Humming Bird)[Hummingbird]
Notes as in 18

No. 21 5/31 (Sita aculeate)[Sitta carolinensis] (Slender-billed Nuthatch) [White-breasted Nuthatch]
Taken on Mt. side from pine tree. Not common
Ruby Creek Ada Co Id.

No. 22 5/31 Cyanospiza [Passerina] amoena Lazuli Bunting
Numerous – Ruby Creek Ada Co Id.

(Eggs) *No. 6 5/31 Falco Sparverius (Sparrow Hawk)* [American Kestrel]
No. in set 4 Size 1.46 – 1.44 – 1.47 – 1.42

Taken from stump of pine tree 20 feet. Nest in cavity 14 inches; in same tree was a nest of Asyndesmus torquatus [Melanerpes lewis; Lewis's Woodpecker]. No nest except bare hole
Ruby Creek Ada Co. Idaho

No. 7 5/31 Cyanospiza [Passerina] amoena Lazuli Bunting
No. in set 4 Nest 4 ft. from ground in small bush.
Ruby Creek Ada Co. Idaho

No. 8 5/31 Accipiter cooperii (Cooper's Hawk)
No. in set 4 Measurement in length 2.08 – 2.08 – 1.96 – 2.16
Nest of sticks lined with fine bark& pine needles. Height 15feet in poplar tree ([blank] *)*

(Birds) *No. 23 6/1 Dendragapus obscurus (Dusky Grouse)*
Notes as in # 19
Ruby Creek Ada Co Idaho

No. 24 6/1 Sunday ------------- Notes as in # 15
Ruby Creek Ada Co. Idaho

(Eggs) *No. 9 6/1 Merula migratoria propinquas* [Turdus migratorius] [American Robin]
Nest taken along creek in slender willows willows 4 ft. from ground. Indications showed that this bird had been building here for a number of seasons as I counted at least 4 old nests. The nest was exactly like the M. migratorius as far as I could see. No. in set 4

(Birds) *No. 25 6/3 Asyndesmus torquatus [Melanerpes lewis] L.* [Lewis's] *Woodpecker*
Notes as in # 10

(Mammals) *No. 5 6/3 (Little Chipmunk)*

Common l of body – 3.25 Tot[al] *length 9 Cir of Body 350 l of tail 4*

No. 6 6/3 (Little Chipmunk)
Notes as in # 5

(Birds) *No. 26 6/4 Asyndesmus torquatus [Melanerpes lewis] L. [Lewis's] Woodpecker*
Notes as in # 10

No. 27 6/4 Dendragapus obscurus (Juv) Sex ?
Flushed mother on Mt. side found only the one chick.
Ruby Creek Ada Co. Idaho

No. 28 6/4 Pica pica hudsonii [Pica hudsonia] ([Black-billed] *Magpie) Juv*
Taken in ravine off Ruby Creek

No. 29 6/4 Pica pica hudsonii [Pica hudsonia] ([Black-billed] *Magpie) Juv*
Notes as in # 28

(Eggs) No. 10 6/5 Colaptes cafer (Red-shafted [Northern] *Flicker)*
No. in set 8; Location – pine stump height 8 ft. well incubated

No. 11 6/5 Icterus bullocki (Bullock's Oriole)
No. in set 6 Description in skin catalogue [See No. 41, 6/11 account]

No. 30 6/6 Pica pica hudsonii [Pica hudsonia] ([Black-billed]*Magpie) Juv*
Notes as in above (28& 29)
This bird is more common in this immediate locality but

I am told by settlers that in certain gulches in the county they are numerous. I presume from this that they occur in communities. I have been watching diligently for some of the (Cyanocitta) types but so far have been disappointed. Bird life in the mountains here is disappointing in one way, that is, on account of the fewness of the varieties. In another place I made a record of Corvus americanus [brachyrhynchos] – having shot one of the only pair seen since leaving Kansas. Hawks are rather rare only about 3 varieties seen thus far& those all occur in Kansas (Accipiter cooperi[i]) (Accipiter velox [striatus]) and (Falco sparverius).

No. 31 6/6 Pyranga [Piranga] ludoviciana (Louisiana [Western] *Tanager)*

Same as in #13 This bird I have not yet located. It is the most brilliant type by far I have found. Following are 3 others of the same type all male birds. They seemed to be in migration and are easily approached. The one taken May 30 (No. 13) had not developed the brilliant red at that time.. There seems to be a varying degree of brilliance in the red head. The song resembled that of (Icterus galbula). (l of body 2 in) Number in flock 6

No. 32 6/6 Pyranga [Piranga] ludoviciana Louisiana [Western] *Tanager*

Notes as in 31

No. 33 6/6 Pyranga [Piranga] ludoviciana Louisiana [Western] *Tanager*

Notes as in 31

No. 34 6/6 Pyranga [Piranga] ludoviciana Louisiana [Western] *Tanager*

Notes as in 31

Ruby Creek Ada Co Idaho

No. 35 6/6 Merula migratoria propinqua [Turdus migratorius][American Robin]
Ruby Creek Ada Co. Idaho

No. 36 6/6 Geothlypis [Oporornis] tolmiei (Reed) [MacGillivray's Warbler]
Saw several took this one only; in deep ravine off Ruby Creek Ada Co. This is the first warbler taken here- hope to get others soon.

(Mammals) *No. 7 6/6 (Pine [Red] Squirrel) [Tamiasciurus hudsonicus]*
Shot by Green from a tall pine tree. The distance must have been 100 yds. Down the ravine. He used a 22 Marlin rifle. These little animals are quite common — making nests of pine needles & bark at various elevations mostly from 10 to 20 ft. high. l of body skinned 6.25 - tot. l 15 inches.

(Birds) *No. 37 6/7 Dendragapus obscurus [Dusky Grouse] (Baby Grouse)*

No. 38 " " " "

No. 39 " " " "

No. 40 " " " "
My dog Jo flushed old hen on Mt. side and we captured these little ones after a hard scramble.

(Mammals) *No. 8 6/7 Mouse*
Taken on Ruby Creek along stream — trapped.
Length 9.2 l of body 2.5 l of tail 275

No. 9 6/7 Mouse
Notes as in # 8

(Eggs) *No. 12 6/10 Stellula calliope (Calliope Hummingbird)*
No. in set 2 - half incubated Position of nest pine tree ([blank])
10 ft. from ground on longitudinal bough.

No. 13 6/10 Accipiter cooperii (Cooper's Hawk)
No. in set 4 Notes and measurements as in # 8

(Birds) No. 41 6/11 Icterus bullocki (Bullock's Oriole)
l of body 2 depth 1.25
This bird was taken in the most remarkable manner, which I shall try to describe here briefly. W. D. Green my assistant and myself had been out this day on a short tramp in the mountains and were returning to camp when something was discovered in a dead pine tree about 40 feet from the ground. We first took it to be a caterpillars nest but upon closer observation saw that it was doubtless some sort of hanging bird nest. To reach the nest by climbing was impossible as it was situated far out upon small branches. After considerable scheming it was decided that we would have to shoot down the nest. Accordingly one of the party prepared to cut away the three small branches supporting the nest by shooting a 22 caliber rifle ball through these while the other stood in readiness to catch the nest when it should fall. The first shot cut one of the twigs squarely in two leaving the remaining two holding the nest. Four shots were fired before the next twig was cut leaving the nest supported by one single twig which was the smallest of the three being hardly larger than a lead pencil. Four shots were fired before this one was cut and the nest dropped into the hands which were waiting to receive it. The question now was, what was in it? The first peek revealed the fact that a live bird was upon the nest or in the nest which is 10 inches deep. The second peek revealed the fact that it was not only a living bird but the adult female sitting as tho nothing had

taken place at all unusual. She was taken from the nest and chloroformed. And now comes a more remarkable thing than either the shooting away of the nest or the capture of the bird alive in this way — in the bottom of the nest were six eggs all unbroken. These eggs are exceedingly delicate, a characteristic of the eggs of the Icteridae group. They had dropped 40 feet. The bird sitting upon them and over them weighed enough to crush a hen's egg. The eggs were found to be well incubated which may explain the behavior of the female in refusing to desert the nest. I regard this as one of the most remarkable collections I have seen considering the fact that the type is rare and the circumstances absolutely unfavorable to the taking of any part of the set in perfect condition.

(Eggs) *No. 14 6/12 Zamelodia melanocephala [Pheucticus melancephalus] (Black-headed Grosbeak)*
No. in set 4 nest 10 ft. high in small Quaking Asp tree Female upon nest and was taken Mores Creek Boise Co. Idaho

(Birds) *No. 42 6/12 Zamelodia melanocephala [Pheucticus melanocephalus] (Black-Headed Grosbeak)*
Taken from nest [see above egg-collection note]
L of body 2.25 Depth of body 1.25 Mores Creek Boise Co. Idaho

No. 43 6/13 Icterus bullocki (Bullock's Oriole)
Taken on Mores Creek Boise Co. Idaho
L of body 2 in depth of body 1.25

No. 44 6/13 Troglodytes aedon aztecus (W. House Wren) [House Wren]
Taken on Ruby Creek Ada Co. Idaho

No. 45 6/13 Troglodytes aedon aztecus (W. House Wren)

[House Wren]
Notes as in # 44

No. 46 Contopus virens [sordidulus] richardsoni [Western Wood Pewee]
Taken on Ruby Creek Ada Co. Idaho

No. 48 6/10 Stellula calliope (Calliope Hummingbird)
Taken with nest & eggs [See egg collection No. 12, on 6/10]
Ruby Creek Ada Co. Idaho Photographed her on nest – camera 2 ½ ft. away.

No. 49 6-14 Dendragapus obscurus (Dusky Grouse)
Notes as before

~~~~~~~~~~~~~~~~~~~~~

It is worth noting that three of the six 1907 taxidermy relics that survived the dismantling of the old Baker Museum and fell into my possession in 1976, were collected during the Baker team's encampment at Robie Creek. On one of these specimens, a male western tanager, the label reads: June 6, Ada Co. Idaho. In Monahan's old field journal it is recorded as one of four tanagers collected that day. Another old museum survivor was a black-headed grosbeak that was collected on June 12, when the Baker Team sallied a little further from their camp into Boise County on Mores Creek. The third surviving specimen is one of the two Lewis's woodpeckers collected at Robie Creek on May 29. Despite the diary record's silence throughout this encampment at Robie Creek, even before my much belated look at the field journals these three bird specimens lent silent testimony to the fact that the Baker Team was collecting specimens during the nearly month-long layover on Robie Creek.

Combining the 26 Robie Creek "observed birds" list with the four additional species collected but not on that list (western wood
~~~~~~~~~~~~~~~~~~~~~

pewee, house wren, western tanager, and Bullock's oriole) the Baker Team encountered 30 bird species during their stay on Robie Creek. This list surpasses the 28 species of my own observations on modern day visits to Robie Creek (see list below). Ten of the 26 bird species were observed but not collected at Robie Creek: turkey vulture, golden eagle, mourning dove, frosted [common] poorwill, least [Empidonax?] flycatcher, American crow, black-capped chickadee, American redstart, western meadowlark, and purple [Cassin's?] finch.

The journal notes reveal several other interesting observations. First, is Monahan's apparent disappointment in the general lack of bird diversity he is finding at Robie Creek. His stated dismay at the paucity of raptor species (he records only American kestrel, Cooper's hawk, sharp-shinned hawk, and golden eagle) is another. And of course there is the remarkable story of the shooting down of the oriole nest with its incubating female sitting tight, still alive, and her six intact eggs when the nest is hand-caught at the end of its gunned-down freefall.

Monahan's journal records provide us a snapshot of late-May to mid June birdlife at Robie Creek. Judging from the egg collections many of the Robie Creek birds were already nesting, yet some, such as the flock of western tanagers and possibly the American redstart were likely passing through in migration. With the exception of dusky grouse, there were more Calliope hummingbirds collected (six) than any other species. Was this a measure of the hummer's abundance in the area or a reflection of their natural tendency in courting and territorial behavior to perch prominently, making them ready targets?

Mammals collected while at Robie Creek were few and fairly vague in their species identity. Since rodents comprise the largest mammal group, it is no surprise that all nine specimens from Robie Creek were of the Order Rodentia. Moreover, considering the unusually high diurnal activity of members in the Sciuridae (Squirrel /Chipmunk) family, neither is it a surprise that seven of these mammals were members of that family. The first four specimens taken were three different kinds of squirrels: "mountain", "pine" and "mountain flying" squirrels. These can be fairly reliably transcribed to Columbian ground squirrel (*Spermophilus columbianus*), red (tree) squirrel, and northern flying squirrel. With the scant information accompanying the two chipmunk records, their identity is more problematic. However, given the elevation and landscape of Robie Creek, they were probably either least [*Tamia minimus*] or yellow pine chipmunk [*T. amoenus*].

Before we follow Monahan and Green into the mountains to the north, let's take a quick tally of wildlife observed on my own century-later visits to the Robie Creek area (mostly in vicinity of the park area).

Birds are still abundant there. In the order they appear in your bird guide, we have observed the following: Canada goose, common merganser, turkey vulture, osprey, golden eagle, red-tailed and Swainson's hawks, American kestrel, California quail, ring-necked pheasant, killdeer, spotted sandpiper, ring-billed gull, rock pigeon, mourning dove, common nighthawk, rufous hummingbird, calliope hummingbird, belted kingfisher, Lewis' woodpecker, northern flicker, red-naped sapsucker, hairy woodpecker, western and eastern kingbirds, black-billed magpie, American crow, common raven, tree, violet-green, cliff and barn swallows, American robin, gray catbird, European starling, cedar waxwing, yellow warbler, western tanager, spotted towhee, lazuli bunting, brown-headed cowbird, Bullock's oriole, pine siskin, house finch and American goldfinch.

We've done no mammal trapping there, however, our visits have produced sightings of chipmunks, muskrats, Columbian ground squirrels and mule deer. Even without fishing for answers, we've seen evidence of sucker and small-mouth bass along the fringes of the open water near the park area.

Whatever herptiles the Baker team may have encountered at this locality would not have included the large bullfrog that we find today. This jumbo species has been wrongly introduced from outside the state, and continues to wreck havoc on Idaho's smaller, native frog species. The magic combo of water and rocks likely supplied Monahan & Green with a variety of snake specimens, if they were interested. Of several possible residents, we've seen both the racer (western yellow-bellied) and the western terrestrial garter snake.

We have no record of the insects noted by the Baker team, even though it was a stated goal to collect them along the way. I'm no entomologist, but with help from Allen Dale and our side-kick butterfly catcher (my grandson) Wesley Dillon, our modern day treks through this area have catalogued some of the more conspicuous butterflies. Cabbage, and marbled whites, Sara's orange-tips, checkerspots, small wood nymph, large wood nymph, pearly crescent, silver spotted skipper, mourning cloak, coral hairstreak, gray hairstreak, ochrea ringlet; tailed, western, and pale swallowtails, satyr anglewing, greater spangled fritillary, three sulfurs (clouded, orange and Queen Alexandra's), blues (acmon, arrowhead, Boisduval's, dotted, Melissa, silver, spring, western tailed) and the red admiral frequent this area.

Besides butterflies, another colorful group of large insects difficult to not notice are the damselflies and dragonflies. Though some are more challenging to catch and identify, we have recorded the emerald spreadwing, blue-eyed darner, white-tail (common or desert?), eight-spotted skimmer, and Pacific spiketail

In a more general category, we've found the Carolina grasshopper, water striders, and large black spider wasps.

Plants (listed by common name in alphabetical order) have included: alder, arrow-leaf balsamroot, bitterbrush, black locust, blazing star, buttercup, ceanothus, cattail, cheat grass, chokecherry, clover, chicory, white clematis, collomia, cottonwood, Douglas-fir, dogwood (red-osier), dogbane, elderberry, evening primrose, yellow salsify, golden current, goldenrod, gumweed, mountain hawthorn, horsetail, indian paintbrush, lupine, milkweed, mountain maple, climbing nightshade, serviceberry, streambank globe mallow, twinberry and willow.

At some point between their original proposal for the expedition and their actual trek the expedition's intended destination changed. Early Baker University announcements of Monahan's 1907-08 leave of absence stated Idaho's Sawtooth Mountains as the destination. Ultimately, however, they struck north from Boise to Payette Lake, instead of heading northeast into the Sawtooth range. There is no explanation offered for this altered objective. Perhaps during their month-long stay in Boise, and after periodic visits with the Bacon family, they were persuaded to change their plans. Whether or not it had to do with better travel routes, better prospects for specimens, or better prospects of having their two acquaintances join up with them at mid-summer, is unknown.

As the Baker Team embarked on their summer expedition in Idaho, our nation was struggling with a crumbling economy. Our banking system was teetering on the brink of collapse somewhat reminiscent of today's debacle. President Teddy Roosevelt was in the process of establishing the national forest reserves, forerunner of the U.S. National Forests. Railroad systems were snaking their way into many parts of the West, and San Francisco was rebuilding from its devastating 1906 earthquake. Later in the summer a lone farmer by the name of H.O. Blanchard of Renton, Washington in protest of high-priced city wholesalers, would drive his horse-drawn wagon to the newly planked Pike Place at the Seattle waterfront and quickly sell out his produce, spurring what would become the famous Pike Place Market. On the local scene, the May 25 *Idaho Statesman* newspaper that reported on the Baker Expedition also carried a story of the Boise City Council's labor dispute with city laborers, mediating an agreement that they work nine hour days at wages raised from $2.40 to 2.50 per day. The State Capitol Building was still under construction having been started in 1905. The Davis family was drawing up papers for the establishment of Boise's crown jewel, Julia Davis Park. Boise was also abuzz in jury selection for the upcoming trial of William D. Hayward, implicated in the Governor Steunenberg assassination. Swan Falls Dam had already been constructed and its electricity was slowly making its way across the Boise Valley. The Diversion Dam on the Boise River was under construction and water would soon be making its way to Deer Flat Reservoir near Nampa. Southern Idaho towns were rising up out of the sagebrush, and scattered agricultural developments were taking root.

·7·

NORTH TO PAYETTE LAKE

(JUNE 22-JULY 5)

On Saturday morning, June 22, 1907, the Baker team broke camp, and loaded up their horse-drawn wagon to begin their journey north to Payette Lake, their summer destination. Around 10:00 A.M. they bid their beloved little Robie Creek camp farewell and struck north toward Idaho City. Despite their fondness for Robie Creek, it must have been a heady experience for the team to finally have their expedition in motion. They rumbled along steadily, covering 10 miles before stopping to make camp for the night.

Little else is reported in the diary that day, except that they had forgotten their hatchet back at Robie Creek. That probably came to light as they set about preparing their evening cooking fire. Not to worry though, as Monahan reports in the diary that evening, "... chopped wood with our feet". We also get the first evening meal "menu" report of one egg each, along with tea, bread and butter.

The following morning, Sunday, prompted a travel decision. The question was whether or not they should travel on the Sabbath? They decided to travel that day, but thereafter rest on Sunday. Because of their affiliation with Baker University, a Methodist school, it is likely that Monahan and Green may have held some religious convictions regarding honoring the Sabbath. There is no mention made of religious issues, except for a later reference to Will Green's brother, Snowden Ray Green, being a "teetotaler". With very few exceptions, throughout the remainder of the expedition, Sundays are mostly quiet days around camp with little if any travel. Additionally, on this subject of the Sabbath, there is no indication that the Baker team ever attended church services, even though there were churches near their eventual summer camp in the Lardo (McCall) area, one of which advertised Methodist-Episcopal services (Callendar 1987). Yet, given the primitive conditions under which the men camped and traveled, Sunday-go-to-meetin' apparel was probably not priority cargo. Laundering their field clothes involved boiling them for a day or so before washing them. Their rustic, expedition lifestyle simply wasn't conducive to attending formal social events.

MONDAY, JUNE 23 – The team began their second day's journey around 9:00 A.M. They might have started sooner except that Bally the old bay horse balked and gave them a little trouble. Soon they were on their way to Idaho City, and found good roads all day. Will Green drove most of that day and Monahan reports taking several fine insects along the way. No clues are provided regarding the identification of the insects. Of course the possibilities are nearly endless, but since their passage follows mostly creek-side trails, I envision the flash of colorful butterflies as well as dazzling iridescence of dragonflies and damselflies.

Their day's travel didn't quite take them into Idaho City. They stopped and camped along a small creek not far from the town. Will cooked supper while Monahan tended to the horses.

Our present day observations along this portion of their trek have recorded: (Birds) Canada goose, turkey vulture, red-tailed hawk, American kestrel, spotted sandpiper, common raven and barn swallow; (Reptiles) wandering garter snake, western fence lizard; (Mammals) northern pocket gopher; (Fish) rainbow trout, squawfish, sucker; (Insects) Western Hairstreak butterfly, stonefly, caddis fly, and carrion beetle; (Plants) (In additional to many of those already listed for the Robie Creek area) duckweed, elodea, fireweed, milkweed, mint, oxeye daisy, quaking aspen, thimbleberry, and wild buckwheat.

MONDAY, JUNE 24, began at the early hour of 5:20 A.M. The Baker team made ready for an early start for Idaho City. Green took the old bay horse, Bally on a quick warm-up ride hoping to avoid a repeat of the previous morning's balking episode. That seemed to do the trick and the men were off without incident and on their way by 6:00 A.M.

Not far up the road, the Baker Team met a miner from California who reminded them of one of their Baldwin acquaintances, Jno. Rhinly. The miner, decked out in white pants, had a drill outfit and was prospecting for gold. They visited briefly and before departing, Monahan gave the miner a package of Curve Cut tobacco, refusing pay from him. This is the first of many instances reflecting Monahan's amiable and engaging personality as the expedition encounters a hodgepodge of frontier individuals along the way. He seemed to flow with a natural generosity.

The Baker team rolled into Idaho City that morning and took time out to write and mail letters to several friends. Monahan describes the place as a typical mining town with all the houses of unpainted lumber. Before leaving town they purchased a new hatchet (replacing the one they had forgotten at Robie Creek) for $1.15.

On the road again, the team began a gradual ascent toward Centerville. They stopped again briefly at noon, possibly for lunch, and then continued on before camping for the day around 2:00 P.M. They were now between Idaho City and Centerville. Here the diary records their observation and collection of their first bluebird – with no indication as to whether it is a mountain bluebird or western bluebird. However, the field journal clarifies this with its logging of a mountain bluebird specimen being taken that day.

Somewhere along the way they must have collected a hummingbird, because the diary records them visiting with a Nebraska sheepherder who traded them six arrowheads for a hummingbird skin. Again, Monahan's friendliness and experience as a taxidermist and biology instructor comes to the fore as he voluntarily instructs the sheepherder in the delicate skinning process.

From Monahan's 1907 Field Journal:

6/24 No. 50 Sialia artica [Sialia currucoides] Rocky Mtn. Bluebird [Mountain Bluebird]
Boise Co. Idaho

TUESDAY, JUNE 25 The Baker team experienced a very cold night in their camp between Idaho City and Centerville. Perhaps because of that they slept late and didn't get on their way until 9:00 A.M. They passed into big timber that morning where Monahan collected a white-headed woodpecker. Below in his field journal notes, Monahan likens the call of this woodpecker to that of the red-bellied woodpecker, a bird he was more familiar with near his Kansas home.

They stopped at noon just out of Centerville, then rolled on through the town around 2:00 P.M. The town they visited was the "old' Centerville town, located a short distance across the valley from its present-day site. Monahan describes the town as "once being a stirring place but now nearly deserted."

They stopped and conversed with an old-timer who was chopping his winter's wood. A resident of the area since 1864, he regaled the Baker team with tidbits of local history. He boasted how he won and lost a fortune, and that people in that area lived so long that many go blind in their old age.

Proceeding on toward Placerville, they saw their first Canada (gray) jays and collected a couple young ones. Though not recorded in the diary, a female mountain bluebird is also collected from among the roadside pines. Within sight of Placerville and earshot of hydraulic mining, the team stopped and camped for the night. Despite the noise of miners, their campsite was deemed a nice place with good water and plenty of grass for the horses. An important and abiding consideration for campsites throughout their travels was the presence of water and grass for the horses. Little else is reported about their livestock food but apparently the horses' diet was sometimes supplemented with oats. The diary ends that day with the reference to a fine bath in a nearby artificial ditch. That night the campers slept warmer than the night before.

~~~~~~~~~~~~~~~~~~~~~

From Monahan's 1907 Field Journal:

[Birds] 6/25 No. 51 Perisoreus canadensis (Canada Jay) [Gray Jay] *Juv.*

Taken near Placerville Idaho Boise co. from a tall pine
This with No. 52 is the first seen thus far

6/25 No. 52 Perisoreus Canadensis (Canada [Gray] *Jay) Juv.*

Notes as in 51

66

6/25 No. 53 Zenopicus [Picoides] albolarvatus (White-headed Woodpecker)

This bird I shot from a pine tree on this date. He was noisy and I thought at once of the Centureus carolineus as his call was very similar. This is the first time I have met this bird in Idaho Taken near Placerville Boise Co.

6/25 No. 54 Sialia articus [currucoides] (Rocky Mtn. Blue-bird) [Mountain Bluebird]

I first met with this bird in Idaho City Boise Co. It seemed to be domesticated thee much as our own Eastern S. sialia; living in boxes & deserted chimneys. Today we saw several pairs as we drove through the pines. Have found no nest yet but hope to soon as I do not think they are hatching yet.

~~~~~~~~~~~~~~~~~~~~~

Modern day excursions through this same area have recorded the following wildlife: red-tailed hawk, osprey, common snipe, mourning dove, Vaux's swift, downy woodpecker, lazuli bunting, western wood pewee, violet-green swallow, western bluebird, chipping sparrow, dark-eyed junco, red-winged blackbird, brown-headed cowbird, Cassin's finch, and pine siskin. We've also seen mule deer and moose.

WEDNESDAY, JUNE 26 Having slept warmer that night, the Baker team was up at 6:00 A.M. Arising early became a common practice for the expedition. Within the diary entries, their arising time is noted on 30 dates. Half of these were actual clock times (5:20 A.M., 6:00 A.M., etc.) and half were noted as either "early" or 'late". On roughly three-fourths of these mornings they were up early. Late sleep-ins were occasionally recorded, but those seemed to occur under peculiar circumstances – when socked in by rain, after sleeping uncomfortably cold at night, or rare instances when one of them was ill. In early fall, several hunting trips for deer and mountain goats got them up around 4:00 – 4:30 A.M. But those were extenuating circumstances. Typically, the Baker team started their days around 6:00 A.M. As close to the 45th Parallel as they were, Idaho's summer solstice brings long days and short nights.

While Will Green prepared breakfast, Monahan went to round up the horses. Old Bally was quickly located but the one-eyed, gimpy-legged horse, Star Crust, couldn't be found. After breakfast both men searched until they located the missing horse, and finally hit the road by 8:00 A.M., headed for nearby Placerville.

On this, their fifth day on the trail, Monahan and Green initiate their quirky use of aliases in the diary. Monahan becomes "Cot Pringle", and Will Green goes by "Arthur Capper". This shift is sudden and without explanation. Nor am I aware of any literary precedence for this moniker use. Yet, as I write this book I find an ironic corollary to this oddity offered up by our own current Idaho Governor, Butch Otter. Turns out that the governor and one of his long time friends, Mike Gwartney, head of the State Department of Administration, oftentimes revert to their own private world of aliases, especially while they played cowboy on Otter's former Eagle, Idaho ranch ,"Lonesome Dove". For many years the two cowboys assumed the Texas Ranger aliases from the Lonesome Dove story; Otter being Captain Call and Gwartney, Gus McCrae. These aliases were so routine that at the Eagle ranch where they practiced their team-roping, two chutes bore the names Capt. Call and Gus (The Idaho Statesman, Feb 10, 2008). It's sort of "grown-ups" playing let's pretend. I suspect these noted examples are just a few of the other probable aliases sported by some of Idaho's colorful cast of public characters. I've even proposed a few appropriate aliases for certain Idaho politicians. But, of course, you'll need to exhume my diary if you ever expect to know any of them.

Their stop at Placerville prompted this assessment "...the best town since we left Boise." While in town they picked up a few supplies. One item, a bell, was probably inspired by their search for their lost horse earlier that morning. Another purchase was a quilt, no doubt to remedy their sleeping problems on cold nights. Other supplies included a pound of butter, bushings for the wagon, plaster of Paris and gum.

A short distance out of Placerville they stopped along Granite Creek and had lunch. On my modern-day expeditions with my students, we likewise stop along this same Granite Creek for lunch to commemorate the Baker lunch hour. Today it remains a green, beautiful area for picnicking, with beaver ponds sometimes in evidence. Here we have observed turkey vultures, a sharp-shinned hawk, red-tailed hawk, American kestrels, gray jay, and American crows. We've also recorded the more colorful belted kingfisher, Lewis's woodpecker, northern flicker, barn swallow, red-breasted and white-breasted nuthatches, ruby-crowned kinglet, American robin; Nashville, yellow-rumped, and Townsend's warblers, black-headed grosbeak, and Cassin's finch.

But life on the 1907 Expedition wasn't entirely idyllic. During their lunch break, the horses promptly ran off after being fed. To add to their annoyances, Green broke his pipe. But these were minor concerns and the team was soon on its way again, ascending the five-mile stretch of road toward the top of Harris Creek summit. Around 6:00 P.M. they reached the summit and camped. Even at this 5,213 foot elevation the afternoon was hot and thick with flies. After turning their horses out to graze, Green and Monahan enjoyed the beautiful view and sought some coolness by eating snow from nearby snow banks and drinking ice cold water.

The diary also records one intriguing wildlife observation at this spot. Even today there remain hillside patches of sagebrush, and here the Baker team encountered two sage grouse, though they were unsuccessful in collecting either. According to Idaho Fish & Game Department Upland Game Biologist, Andy Ogden, there are no longer any greater sage grouse inhabiting that area. Given this grouse species' propensity for migration – sometimes as much as 50 miles – and today's severe declines in population numbers, this record stands as a dark reminder of the current extent of sage grouse extirpation – their total loss from some areas.

The diary notes that no specimens were taken that day. Minus their would-be sage grouse collection, supper that night remained the simple fare of toast, tea and bacon.

THURSDAY, JUNE 27 The new quilt did the trick for granting the men a good night's sleep. That is, until about 5:00 A.M. when their two old horses wandered into camp and awakened them. Despite this unexpected equine wake-up call, the team slept on until 6:00 A.M.; they were up and off by 7:30. On this day the team rumbled down the rocky, windy road from Harris Creek Summit to the bottomland of the Payette River near Horseshoe Bend. Monahan describes their descent as:

"First two hours looped the loop then road went straight down. Had to fix brake with old shoes, tied rocks on behind."

Around noon the team camped for lunch on a little rise overlooking a ranch. In a nearby creek they fished for trout, where Monahan succeeded in catching one. Here, and later on in the expeditions story, I get the impression that the older leader was the better angler of the two men. But as the summer progresses, numerous fishing opportunities serve to hone Will Green's skills as well.

Continuing downward, they jostled lower and lower along the deep ravine road cleaving the bare, elephant-head shaped hills. The road, very rocky and rough, dislodged a bolt from a front spring, which they were equipped to replace. Passing by farmland, Green evidently took advantage of some of the local "tobacco" stock for his pipe, for the diary states that he smoked timothy and alfalfa. Monahan, now suffering from a sore tongue, didn't join in.

At this juncture they also shot six mourning doves for their young hunting dog, Jo. Here too, they shot a "Red-head". I have long pondered (for 36 years – from my first reading of the diary in 1973 until this writing in 2009) the exact identity of this "Red-head" species. This is the first of four diary entries that allude to this bird, either as a specimen or the taking of its nest. Granted, there is a redhead duck but in most of these diary–recorded red-head encounters, neither the habitat nor setting seems appropriate for the redhead duck. And remember, 1907 was still, relatively, the dark-ages of avian nomenclature – i.e. naming of birds. My quandary: if Red-head (more often inscribed "redhead" in the diary) was not the waterfowl species, what was Monahan actually referring to? I've considered other logical candidates for this mystery bird. Those include the male gender of either the western tanager, Cassin's finch, or one of the red-headed woodpeckers. Yet, the *diary* account offers no conclusive clues. The answer, I think, finally came with my long belated 2009 discovery of Monahan's old field journals; in particular, within the field journal entry below. Note three things: there is no species name offered, only the words, "Not yet identified". Secondly, this specimen was a female, which among birds are customarily less descript in color and identifying features than male birds. Thirdly, and importantly this specimen is referred back to # 31 which was a male "Louisiana Tanger" a.k.a. western tanager. Added support of this rationale lies also in the other 1907 common name for the western tanager (in addition to Louisiana tanager): the "crimson-headed tanager", which loosely and quickly equates to "red-head".

Note too, that the collection date for this implied western tanager is June 26. This is surely an error in date as the diary specifically states on June 26 that no specimens were collected. Additionally, the diary puts the redhead collection at the same time as the taking of the Brewer's blackbird nest, and both diary and field journals place the collection sites for both along the Payette River which they didn't reach until June 27. The diary reports that their camp that evening was on the banks of the Payette River above the town of Horseshoe Bend.

Though there is no mention of this in the diary, toll roads were somewhat common in the area at that time. It was one of the few means of financing the building and maintenance of early roads. Today, at the base of the road leading down from Harris Creek summit to Horseshoe Bend, there stands a monument to a toll gate that resided there in 1907. Very probably the Baker team passed through it, paying their toll as they entered the little valley.

~~~~~~~~~~~~~~~~~~~~~

[Birds] *6/26* [*6/27?*] *No. 55 in # 31 Not yet identified*
Taken on the Payette R. Boise Co. Id.

[Eggs] *6/27 No. 15 Scolecophagus [Euphagus] cyanocephalus Brewer's Blackbird*
No. in set 4 Nest in Haw[thorn] *bush beside road. Taken on Payette R.*
Boise Co. Id.

~~~~~~~~~~~~~~~~~~~~~

FRIDAY, JUNE 28 Up at 6:00 A.M. the team was on the road by 7:00. They stopped at Horseshoe Bend, at that time described simply as a store with a post office inside. The diary says nothing about purchases at the store. Perhaps they mailed letters or simply stopped in to visit and inquire about the road conditions that lay ahead of them. Monahan already alludes to the ominously steep stretch of road ahead of them; up and out of the canyon,'1 ½ miles straight up' near the Dry Buck summit.

The canyon bottom of that area fairly scorched the Baker team with their hottest day yet. By noon the heat and the mosquitoes were so bad that they decided to camp for the day and start out early the next morning. They didn't relish the idea of tackling the steep Dry Buck grade in such heat. Even their puppy complained from the heat and fatigue. Though the young bird dog had made his first real point on a dove that morning – "beautiful form", says Monahan – by the time they stopped at noon, he was howling in misery, and retreated wearily to his bed beneath the seat of the wagon.

Trying to make the best of the heat and mosquitoes, the two men sought solace in fishing for awhile along the Payette River. But even in the hottest part of the day, the mosquitoes, tiny but ferocious, were unbearable. Monahan and Green were soon driven back to camp, where they built a circle of smudges around them to fend off the insects. Not surprisingly, under such taxing conditions no scientific specimens were taken that day.

The Payette River along which they are camped is described as clear, cold, rapid, and reported by locals to be treacherous. The river's deadly reputation was legitimate. Earlier that summer, on June 1, 1907, a crew of seven loggers were drowned on the annual log drive down the South Fork of the Payette. During the previous two to three decades several loggers had perished on similar runs on the North Fork of the Payette; one above Horseshoe Bend. Those dangerous log drives on the Payette and other river systems ended soon after the railroads extended northward into the mountain country between 1912 and 1914. Today, a century later, it is outdoor recreationists, mostly kayakers, good and strong as well as the naïve, who perish on those same waters nearly every year. The Payette River remains cold and clear enough today to entice river rafting tours in its calmer waters, and yet treacherous enough in its white waters to challenge the best of kayakers.

As the men settled into the smoke-smudged evening, Green spent some time oiling the "old Science Car" in preparation for an early morning assault on Dry Buck Mountain.

SATURDAY, JUNE 29 The expedition team rolled out of camp at 7:00 A.M. Their stop back by the Horseshoe Bend store seems precautionary and betrays the concerns weighing heavily on the men's minds that morning. They purchased a couple bolts for the wagon and oats for the horses. I think they knew that ahead of them lay the most challenging travel day yet. At the store they inquired about the Dry Buck road and were informed that it was the steepest stretch of regular road in the state. With that, the team started up the road prepared to conquer the hill.

Curiously, there is no mention of crossing the Payette River anywhere in the area of Horseshoe Bend. And there is no mention of Banks. Whether the Baker team crossed the Payette by bridge or ferry, Idaho historian, Jerry Branson, feels their probable ascent of Dry Buck mountain was through the valley near present-day Gardena.

From the onset the Dry Buck road zig-zagged steeply upward. Soon they came to such a dangerous section of the road that they lightened their wagon by unloading half of their belongings in order to safely proceed on. Green drove while Monahan pushed from behind as best he could. The road was so steep and the footing so rough, it was all Monahan could do to walk, much less push. Nonetheless, by plodding ahead 50 to 100 feet and then resting, the horses performed admirably. Steadily, they made their way toward the top. At times they encountered sandy washouts in the road. Other places were rutted and rocky. Monahan carried a box to block the wheels but even with that he sometimes had to stand on the spokes to prevent the wagon from rolling backwards. In one or two places their progress was outright scary.

Finally the wagon reached the top and the men unhitched the horses. Taking the tent and tarp and horses back down to their jettisoned gear, they were able to load up everything and make it back to the top in one trip. On top of old Bally's load was an insect net, and across Star Crust's pack was a pair of gum boots. The uphill haul wasn't without incident, however. Although Bally was a jewel and packed two-thirds of the load up the hill without so much as a groan, old Star Crust kicked off his load at one point and had to be re-packed and babied the rest of the way up the hill. Their strenuous, successful ascension to the summit of Dry Buck Mountain took two hours. The men felt great relief at having the ordeal behind them, and without serious incident.

The exposed crown of Dry Buck was mostly open hillside, so the team drove on until 2:00 P.M. before finding water for their horses, which hadn't drunk since morning. An hour further along, they also found adequate shade, grass and water for a suitable camp site. For the first time since leaving Robie Creek, they set up their tent and made ready to spend the next day, Sunday, resting and relaxing. As required periodically, they ran their wagon into the creek to soak its wheels. Monahan signs off the diary by noting that even though that day was not so hot, the nights were getting warmer. Having been another horrendous travel day, for the second consecutive day there is no mention of collected specimens.

Nearly a century later, with tamer road conditions and powered vehicles, our student expeditions have enjoyed the luxury of more relaxed wildlife sightings along the slopes and summit of Dry Buck Mountain.

We've observed great blue heron, red-tailed hawk, golden eagle, northern harrier, American kestrel, California quail, mourning dove, Calliope hummingbird, Lewis's woodpecker, northern flicker, olive-sided flycatcher, western kingbird, Cassin's vireo, Steller's jay, common raven, mountain bluebird, American robin, cedar waxwing, yellow-rumped, yellow, and MacGillivray's warblers, western tanager, spotted towhee, chipping sparrow, lazuli bunting, red-winged blackbird, pine siskin, and American goldfinch.

Other wildlife that we've seen on Dry Buck includes mule deer, coyote, beaver, northern pocket gopher, Columbian ground squirrel, chipmunk, and signs of black bear (tracks), Boreal toads, and twelve-spotted skimmer dragonfly. Wildflowers dress the roadside landscape in colorful profusion amid scattered ponderosa and lodgepole pines.

Brushy ravines were clogged with chokecherry, snowberry, elderberry, ninebark and golden current. Along the way one finds both *Clarkias* (including my favorite, "ragged robin"), Sitka (red & yellow) columbine, western coneflower, dogbane, , California false hellebore, common hound's tongue, lupine, milkweed, western monkshood, moth mullein, mountain maple, pearly everlasting, *Penstemon, Potentilla,* scorpion weed, self-heal mint, snowbush, wild geranium, wild onion, and yarrow. Cavity-riddled snags offer stately punctuation to the timbered landscape. Occasional aspens shimmer in jittery contrast to sedate pines.

SUNDAY, JUNE 30 As agreed upon a week earlier, when they elected to travel on the previous Sunday, this Sabbath, and every subsequent one, was for rest and relaxation around camp. That morning they arose late and enjoyed a sumptuous breakfast prepared by Green. The breakfast menu consisted of Oatmeal, beans, biscuits, coffee, prunes, rye bread, and sorghum.

Perhaps to honor the Sabbath or just for the change of a non-travel day, the two men donned their new khaki pants and shirts. One catches a glimpse of the men's pioneer resourcefulness as they gave each other their first Idaho haircuts, and Green hob-nailed and half-soled his boots. Noticing their rye bread was beginning to mold, they took the precaution of re-baking it. They also cooked up enough peaches to last them for the next two to three days. Later, they pulled the old wagon out of the stream and began repacking it for the next day's travel.

The diary record of the day's end reflects a sense of the mountain quietude and idyllic setting. Monahan sits on a log, smoking and listening to the bell on one of the old horses. In the distance he spots a ruffed grouse flying across a clearing. The puppy has already retired to his bed beneath the wagon seat. And in the diary, Monahan pencils a farewell salute to the month of June with hopes of reaching Payette Lake by the following Sunday.

MONDAY, JULY 1 After their much needed day of rest, the men arose early enough the next morning to be on the road by 6:40.A.M. Good roads would grant the team a banner day of travel, covering about15 miles. Higher up, dense forest bordered much of the road but soon they descended into the lush, broken expanses of beautiful High Valley. Though Monahan was feeling poorly, coming down with tonsillitis, Green took 10 specimens that day. Such a bountiful take on specimens in this sector of their travels isn't surprising. Even today this portion of the expedition's route – from Dry Buck Mountain north through high Valley – is a fertile area for wildlife. This valley remains one of my favorite Shangri-La places to go on a look-about for wildlife.

According to the diary, among the 1907 specimens taken that day were a pileated woodpecker, two red-shafted (northern) flickers, "a strange one not yet identified", four killdeer, and two red heads. Again the ambiguous "red head" crops up among the take. The "strange one" not yet identified, but listed with the woodpeckers, may have been a Williamson's sapsucker, which is the only specimen recorded in Monahan's field journals for that date. They had already noted, hence recognized, the Lewis's woodpecker and red-shafted flicker. On the other hand, Williamson's sapsucker has one of the most restricted western ranges, and would be least familiar to the team from Kansas. Our present day treks through High Valley still find this rare sapsucker nesting there.

Traveling the length of High Valley that day, the Baker team camped in a beautiful meadow with a small trout brook so clear they could see the fish. Once again Green tries his hand at fishing, but again remained unsuccessful. As Monahan notes in the diary: "Not on to it yet."

It is curious that of the 10 specimens collected on this date only the probable "strange woodpecker" (Williamson's sapsucker) is logged in the field journals. It is noteworthy that Green did all the collecting, and I wonder if the fact that Monahan was sick with tonsillitis hindered his work that day to retaining only the sole new specimen? In fact even the journal entry seems in error as to the location of the collection site. The expedition was well into High Valley that day – a full day from the Payette River before and after High Valley.

~~~~~~~~~~~~~~~~~~~~~

(Birds) *7/1 No. 56 Sphyropicus [Sphyrapicus] thyroideus* 77
(Red-throated Woodpecker) [Williamson's Sapsucker]
Taken from Pine tree Not at all shy.
Boise Co. on Payette River

(Mammals) *7/1 No. 10 Mouse*
Notes as in # 8

~~~~~~~~~~~~~~~~~~~~~

Present day summer treks through High Valley never fail to reward the wildlife watcher. In addition to the Williamson's Sapsucker mentioned above, our many trips through the valley have found the following:

Canada goose, common merganser, ring-necked duck, American widgeon, mallard, cinnamon teal, green-winged, and blue-winged teal, common goldeneye, American coot, greater sandhill crane Virginia rail, great blue heron, killdeer, Wilson's snipe, spotted sandpiper, upland sandpiper, long-billed curlew, mourning dove, northern saw-whet owl, common nighthawk, common poorwill, Vaux's swift, turkey vulture, American kestrel, prairie falcon, bald eagle, osprey, red-tailed hawk, northern harrier, northern goshawk, Cooper's hawk, sharp-shinned hawk, wild turkey, ruffed grouse, horned lark; violet-green, tree, barn and cliff swallows, common raven, Clark's nutcracker, black-capped chickadee, mountain chickadee, red-breasted nuthatch, brown creeper, house wren, western and mountain bluebirds, Townsend's solitaire, American robin, cedar waxwing, European starling, northern flicker, red-naped sapsucker, hairy woodpecker and pileated woodpecker, olive-sided flycatcher, western kingbird, Cassin's and red-eyed vireos; yellow, yellow-rumped, and Macgillivray's warblers, bobolink, western meadowlark, Brewer's, red-winged and yellow-headed blackbirds, western tanager, Cassin's finch, pine siskin, American goldfinch, red crossbill, spotted towhee, savannah sparrow, dark-eyed junco (Oregon), chipping and song sparrows.

Because of their extreme rarity in Idaho and elsewhere, it is worth noting that in the early 1980's we found evidence of at least three nesting pairs of upland sandpipers in High Valley. More recent sightings are few and far between. As is the case in many other parts of Idaho, flocks of introduced wild turkey have only recently established themselves in the valley.

Mammal observations are fewer and include beaver, northern pocket gopher, badger, Rocky Mountain elk, moose, white-tailed deer, coyote, and Columbian ground squirrel.

TUESDAY, JULY 2 The Baker team was late getting up and slow getting around. No doubt Monahan's tonsillitis had something to do with that – in his diary he reports, "…I am feeling tough, throat getting worse all the time." To add to their woes, they were barely underway when one of the wagon's front springs broke. Ever resourceful, the men mended it with a strip of iron and wire. It worked so well that Green jests about breaking all the others and fixing them the same way.

Despite this one-hour delay they rumbled their way out of High Valley and dropped back down to the Payette River to the little town of Smith's Ferry. This ferry had been in operation since 1887, providing a popular crossing for the people, wagons and livestock coming down from the High Valley area. In 1891 it was taken over by Jim Smith, giving rise to its permanent name of Smith's Ferry. The ferry boat operated another 12 years after the Monahan Expedition, when it was finally replaced by a bridge built in 1919 by a fledgling company by the name of Morrison-Knudsen (Ingraham 1992). After crossing the river on the ferry at the cost of 75 cents, they stopped at the little store and replenished some of their provisions. Monahan's diary records their purchases as 25 cents for a dozen eggs, and $1.50 for a bushel of oats for the horses.

Today, as we follow the old road across the bridge at Smith Ferry and drive up through the low timbered hills to Round Valley, we are likely following the same trail traveled by the Baker team as they rolled on to Thrend's Station. No longer in existence, this little town drew the favorable comment in Monahan's diary: "...good little town in a beautiful valley." There they stopped and wrote more cards to friends; possibly because of the availability of mail pick-up.

Like the good progress they had made the previous day through High Valley, this lower end of Long Valley afforded them the best day's travel yet – 18 miles – before they finally stopped at 8:30 P.M., finding a camping spot they liked. They are 35 miles from the Payette Lakes, when a weary and somewhat ill Monahan signs off for the day, wishing they were already at the lake and set up in a permanent camp.

Though the diary records nothing of specimens being collected that day, night time brought an unexpected rain and a wildlife surprise: "...heard wolves all night in the mountains." By 1907 Long Valley's livestock industry had not yet burgeoned to the point of spurring mass predator eradication programs that would soon sweep much of the West. By the mid-1900's the purge of large predators like the wolf, mountain lion and grizzly bear would be nearly complete. So thorough was this extirpation of large predators that it required the Endangered Species Act of 1973 to eventually restore viable populations of both the gray wolf and the grizzly. Today both species have recovered enough in numbers to place them under controversial considerations of delisting.

Prior to the very successful 1995 gray wolf re-introduction, their numbers in Idaho were scant, augmented mainly by a few stragglers drifting down from Canada along the western fringes of the Rocky Mountains. However, once the moribund wolf population was resuscitated by the re-introduction project, Idaho's wolf numbers took off and increased to an estimated 285 in 2002, 650 in 2006, and, by the time of the Federal delisting in 2008, the state was thought to have roughly 788 free-ranging wolves. In the fall of 2009, amid controversy and threatened lawsuits, Idaho opened its first hunting season on the gray wolf with a state quota of 220 allowed wolf kills. Last I read there were over 24,000 wolf tags sold, most selling for $11.50. Idaho is divided into 12 different wolf hunting zones each having specific limits on the number of wolf kills allowed within. The McCall-Weiser zone, which includes the Baker Team's 1907 expedition route, allows 15 wolves to be harvested. By November 9, 2009 the 15 wolf quota was filled.

Monahan was no stranger to wolves. A 1908 Baker University newsletter reports that just three months prior to leaving Kansas for the Idaho expedition, he and a colleague staged a successful wolf hunt back in his home state.

WEDNESDAY, JULY 3 Despite the unexpected night of rain, which surely didn't help Monahan's tonsillitis any, the team rolled out at about their usual time that morning and found good roads as they pressed northward into Long Valley. Well named, Long Valley extends some 60 miles from south to north, and for most of its length, averages about eight miles wide. Tall, dark timber swatches cris-crossed the valley floor, following the gentle topography and network of streams. Well into the valley by now, as they rolled northward they were in sight of snow-capped peaks on either side.

Established homesteads throughout the valley had sprouted a mosaic of small settlements, most springing up around farmed plots, giving rise to a scattering of little towns. Some of these towns along Monahan's Expedition route were VanWick, Thunder City, Crawford, Center, Roseberry, Manning, MacGregor, Alpha, Elo, and Lardo (McCall). By 1907 the valley's population was an estimated 3000.

Making good time, the expedition wagon rumbled through the little towns of Crawford, Center and Roseberry. In the latter town they stopped and purchased feed and a loaf of bread, no doubt at the same General Store that still stands today across from the museum. In 1907 Roseberry, newly incorporated, was one of the largest towns in the valley. According to the January 31, 1907 *Long Valley Advocate,* the town then boasted a hardware store, general merchandise store, a confectionery store, I.O.O.F. hall, two churches, a two-room schoolhouse, a post office, a nice hotel and seven residences. Additionally, there were several structures under construction, including a creamery, sawmill building, one store and two more residences. The Roseberry Hotel was an impressive, two-story, 23-room octagon-shaped landmark. Later in 1907 the area's first newspaper, the *Long Valley Advocate* relocated from Lardo to Roseberry.

Knowing that their destination, Payette Lake, was now not far off, the Baker team drove late that day, finally stopping and camping near a ranch. To their pleasant surprise the rancher hailed from their home state of Kansas. I rather imagine the men enjoyed a warm and cordial visit, since the rancher was formerly of Lyon County, less than 50 miles from Baker University's hometown of Baldwin. In the course of their visit the Baker team came away with some fresh eggs and milk. The other good news – the rancher informed them that they were only 11 miles from the lakes.

Calling it a day, Monahan salinated his sore mouth and throat with a dose of hometown medicine and looked forward to the next day's early departure, hoping to arrive in Lardo by noon.

The sole specimen recorded in Monahan's field journals for July 3 was the following:

> (Mammal) *7/3* *No. 12* *(Little) Chipmunk*
> *This little fellow is abundant everywhere, living in old stumps, trees, under rocks, etc. They are a pest to campers, coming in at night to steal away the victuals. They have a funny chatter when molested and are absolutely without fear of man or "other beasts".*

THURSDAY, JULY 4 Despite feeling "...mighty bum. Throat sore and mouth horrible", Monahan arose early, around 5:30 A.M. On this morn of the 131st. birthday of our nation's independence, Long Valley was dismally cold, wet and socked-in with fog. The Baker team was on its way by 7:00 A.M., undaunted and eager to get to the lake, slogging along the muddy road. A mid-morning burst of sunshine lifted their spirits, and even a terrible mud-hole failed to bog them down. By this time it was growing apparent that many of the valley's scattered residents were also on the move, all headed to Lardo for the valley's Fourth of July celebration.

The Baker team rolled into the town just before noon and found a temporary camp spot along the lake shore, weary but relieved to have finally arrived at their expedition's destination. In his diary Monahan reflects on their good fortune to have completed their journey without serious incident. They had traveled nearly two weeks, covering 150 miles of rough roads; some over extremely rugged mountain stretches – all with an inferior team of horses pulling a wagon – yet they had, at last, arrived safely.

After setting up camp and eating lunch, the refreshed Monahan and Green took advantage of a ready-made opportunity to celebrate their safe arrival at the lakes. They indulged in the luxury of attending the community Fourth of July festivities. Lardo was apparently still the predominant center of community activity, although nearby McCall was rapidly emerging. In 1907 McCall already boasted two general stores, a butcher shop, a large amusement hall, a blacksmith shop and delivery barn (Ingram *1992*) .

What manner of mountain town entertainment prevailed on the Fourth of July in 1907? The diary records nothing of pyrotechnic fireworks or parades. By today's standards, recreational activities of that time and place were relatively simple and practical. Nonetheless, it was obviously a time of community fun and revelry, and Monahan records in his diary that he and Green very much enjoyed the pony races, foot races and ball games. Although neither the "ball games" nor their particular teams are specifically identified in the diary account, they were most likely alluding to baseball.

Later in July of 1907 the *Long Valley Advocate* newspaper reported that Roseberry 'again' played Thunder City (near today's city of Cascade) in baseball. Perhaps these teams were among the contestants on that Fourth of July celebration.

Nor was the day without some notice of desired specimens to be collected. Monahan makes special mention of their sighting of mallards in the valley that day. They must have made inquiry to the locals about the ducks because he further notes that they had been informed of the birds nesting in the valley. That was good news to Monahan, who journals: "We hope to get some young and possibly the eggs." As ubiquitous as mallards are today, and I presume were back then, I'm a bit surprised how notably this duck piqued the interest of the Baker team. The only collected specimen recorded for the day is found among the mammal records.

7/4 No. 11 Mouse
Notes as in # 8.

FRIDAY, JULY 5 The Baker team rose and greeted their first full day at their chosen destination. Greater Payette Lake was now their focal point. After oiling their guns, and readying their gear, the men relocated their camp further up the peninsula and established what was to be their permanent summer campsite. Still close to the lake, this location afforded them better grass for the horses and closer access to marshes and wilder stretches of timberland inhabited by some of the wildlife they planned to collect.

Though Monahan was still not fully recovered from his bout with tonsillitis, after setting up camp, and eager to scout out some of their new surroundings, he and Green couldn't resist taking an exploratory hike. Within a mile of camp they found a marsh that promised some good water bird hunting, and they laid plans accordingly for the next day. On their return to camp Monahan collected one of the "mystery birds" of the expedition: "A female pheasant ...a nice specimen," he states in his diary. Was he referring to the Chinese ring-necked pheasant? The official date of first introduction of this exotic game bird into Idaho is given as 1903 (near Kamiah, Idaho)(Burleigh1972). Had other, later introductions brought the bird into Long Valley? History is unclear as to whether they had by that time been released into Long Valley. And even if so, the area where Monahan shot the bird (roughly present day Ponderosa State Park) doesn't strike one as suitable habitat for the ring-necked pheasant.

Back in the 1970's when I first began scrutinizing Monahan's diary for any records of birds they encountered, this particular bird presented a real identity puzzle. My mentor and Ph.D. advisor, Dr. Chuck Trost, judged this "pheasant" acquisition to be that of a sharp-tailed grouse. That is how we listed it in our 1977 publication of the 1907 expedition's bird list. I confess to niggling doubts about our call, given its preferred habitat of open prairie land or arid grasslands. Such a homeland doesn't sound much like the Payette Lake peninsula to me. Still, I concurred with the greater wisdom of my friend Chuck Trost on this matter.

Perhaps this determination was justified. After all, a reading of the next day's diary entry only adds to the "pheasant" confusion when Monahan records how he prepared the "grouse" skin that Will Green shot the day before and further describes how it tasted better than the large dusky grouse.

Thus went our debate and rationale for 35 years with only the published diary for guidance. I resigned the ultimate truth to the ghosts of the 1907 expedition. Without museum records or some other specimen documentation, only Monahan and Green knew the facts of the diary's "pheasant". Then, within the recently and miraculously discovered Monahan field journal is the following notation:

> (Birds) *7/5 No. 57 Bonasa umbella [umbellus] Ruffed Grouse or "Pheasant"*
> *Taken from a fallen tree in deep pine woods. Near Lake Payette*

Among synonyms listed for the ruffed grouse in Elliot Coues' 1903 bird guide are "Drumming Pheasant" and "Mountain Pheasant". That may account for Monahan's placement of quotation marks around the name Pheasant as a second choice of name used in this journal notation. Too, it helps resolve the whole mystery that long revolved around the identity of this specimen. Case closed.

Settled into their new campsite, which would become their home for the remainder of the summer, the Baker team enjoyed their evening repast of tea, cold steak, fried potatoes, bread, butter and peaches. The men probably felt they had truly arrived. They had successfully completed their travel chapter, which had begun way back on May 12, when they left Kansas. Now, 53 days later, comfortably ensconced in their summer base camp, they could concentrate fully on their mission of collecting specimens for the Baker University Museum.

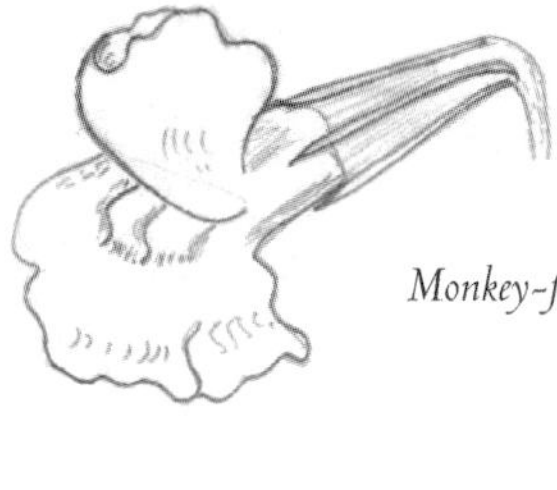

Monkey-flower

·8·

SUMMER CAMP AT THE LAKESIDE

(JULY 6–JULY 23)

SATURDAY, JULY 6 The first couple of days at their new campsite on the peninsula were quiet, spent mostly lounging around camp, relaxing. Their nearly two-week trek from Robie Creek to Payette Lake had no doubt exacted a physical drain on both men; Monahan, especially, because he was still fighting his tonsillitis. Also, the long trek had accrued some domestic chores that needed attending, not the least of which was washing clothes. Now fully positioned for their commissioned task of collecting specimens, they also needed to better equip themselves for navigating the area marshes and lakeshore in search of bird specimens. To that end, Will Green went into Lardo to rustle-up a boat and some needed food items. Monahan stayed in camp where he rested, medicated himself, and *"boiled some dirty clothes"*.

While resting and puttering around their new camp, Monahan also prepared the skin of the "mystery" grouse that Will had shot the day before. Not surprisingly, this specimen served a double harvest – skin for the scientific collection and the flesh washed and saved for supper.

After dining on grouse fare that evening, the men took their new boat out for a look-see at a nearby swamp. Since arriving at the lakes, the team had been on the lookout for mallard specimens, and thought surely they would find them with ducklings here in this slough. However, they were disappointed to find only nesting red-winged blackbirds.

Monahan was faring much better now in his bout with tonsillitis. With hopes of full recovery and venturing further afield the following Monday, before retiring that night the men brought the horses into camp and treated them to some salt. The team would rest on the morrow, the Sabbath.

~~~~~~~~~~~~~~~~~~~~~~

Though the diary says nothing of specimens collected that day, Monahan's bird journal records the following:

(Birds) 7/6 *No. 58* *Loxia curvirostra americana (Am. Crossbill)* [Red Crossbill]
Taken on Lake shore, Payette Lake – small flock seemed to be feeding.

7/6 *No. 59* *L. curvirostra americana (Am. Crossbill)* [Red Crossbill]
Notes as in "58"

7/6 *No. 60* *L. curvirostra americana (Am. Crossbill)* [Red Crossbill]
Notes as in # 58

7/6 *No. 61* *Sphyropicus [Sphyrapicus] thyroideus (Red-throated Woodpecker)* [Williamson's Sapsucker]
Notes as in No. 56

7/6 *No. 62* *Bonasa umbella [umbellus] (Ruffed Grouse)*
Shot from log on shore of Payette lake

7/6 *No. 89* *Cyanocitta stelleri (Steller's Jay)*

~~~~~~~~~~~~~~~~~~~~~~

This jay specimen No. 89 is one of the "surviving seven" Baker Museum specimens sent to me in 1976. Today, after 102 years of wear and tear, this dilapidated jay specimen has relinquished much of its saucy air and cerulean beauty. It's missing one wing and the loose attachment of its head renders it a "bobble-head". The old printed label glued to its stand is dated July 11, 1907, instead of the above journal date. This kind of discrepancy between Monahan's diary account and his field journal recordings occasionally occurs. Whether this results from skipped days of diary entries and later catch-up attempts with inaccurate memory of events, or perhaps the journaling of specimens occurred at later stages of preserving and working the skins for storage.

SUNDAY, JULY 7 On this Sunday the men did rest. They read some, fished a little (without success) and Monahan reported feeling much improved. He busied himself preparing the three red crossbill skins that Green shot along the lake shore the day before. These were Monahan's first ever crossbill specimens and he seemed pleased with them.

On two of my own present-day retracings of the Baker Expedition, I've likewise encountered crossbills in different summers at different locations. The first, I'll never forget, because of the unique sprinkling of pine cone debris that showered down upon us as we stood mid-road along the southeast fringe of High Valley. We had just descended into the valley from Dry Buck Mountain when we spotted a flock of birds alight in the crown of a stately pine shading the road. We stopped, piled out of the vans and began searching the trees through binoculars. We quickly found a sizable flock of red crossbills busily at work on overhead cones plying their uniquely criss-crossed beaks, deftly removing pine nuts, and shucking their feathery sheaths to flutter ground-ward upon us.

Our other encounter with crossbills was east of McCall on a different "Expeditions" campout. After retracing the Baker trek into Long Valley, and finding the original Monahan's campsite area – in vicinity of present day Ponderosa State Park –crowded with people and activity, we chose an alternate site near the Brundage Mountain turnoff where I had once camped and netted forest bats at night. The next morning we rose early and mist-netted and banded birds near our camp. A female red crossbill was netted there along with a Swainson's thrush, chipping sparrow and small (*Empidonax*) flycatcher. The crossbill was a particularly rare catch. Common in much of the western coniferous forests, yet the crossbills' erratic wanderings, following cone crops, added to their ability to breed at any time of the year, render them a rather unpredictable find.

MONDAY, JULY 8 The high mountain air at Payette Lake was visiting a noticeable chill on the Baker team. It was cool enough that Monahan notes their donning of heavier clothing. And, speaking of clothes, it was wash day around camp. Having sufficiently boiled their soiled clothes for two days, it was time to scrub them clean and hang them out to dry.

With the laundry done, the men went into nearby Lardo and picked up some long awaited mail at the post office. This was probably the first day that the post office was open for business since they arrived on July 4th.

Both men were glad to receive news from home – their first in three weeks. Apparently, they were chided for having sent home only postcards instead of letters, but the two enjoyed the fresh correspondence from various friends. One of the letters was from Will's brother Ray Green, a first year high school principle back in Kansas, who reported plans to come out and join them in late August. A piece of less exciting news was a recent copy of an underground student "scandal sheet", *Quid Nienc*, from back on the Baker University Campus. Though thoroughly disgusted by "...the vile sheet...", Monahan was relieved that at least they had not been personally targeted by the printed attacks.

Later in the day, Monahan again tried his hand at fishing, and this time was successful. His hefty catch, three 13 to 15 inch trout, were more than they could eat that night for supper. Rain settled in upon them that evening and the men retreated under cover and were thankful for a good tent and full bellies.

TUESDAY, JULY 9 - FRIDAY, JULY 12TH The rest of the week was spent exploring their new surroundings, especially the peninsula area extending to the north. On Tuesday the two men caught a mess of white fish. The only other "take" that day was a male Steller's jay.

~~~~~~~~~~~~~~~~~~~~

(Birds) 7/9 *No. 90 Cyanocitta stelleri (Steller's Jay)*

Wednesday included another Brewer's Blackbird specimen, their first common merganser and the collection of a set of common nighthawk eggs.

(Birds) 7/10 *No. 91 Scoecophagus* [Euphagus] cyanocephalus *(Brewer's Blackbird)*
Common throughout the state.

7/10 No. 92 Mergus americanus [merganser](American [Common] *Merganser*

(Eggs) 7/10 *No.16 Chordeiles virginianus [minor] W.* [Common] *Nighthawk*
No. in set 2; loc[ated] on ground rocky point on Lake shore Payette Lake, Boise Co. Ida Shot female – eggs slightly incubated

~~~~~~~~~~~~~~~~~~~~

Although Monahan's notes that he shot the female nighthawk, she does not appear anywhere else in his field notes among specimen records. Perhaps Monahan considered the oology record sufficient in itself for logging the taking of the incubating bird. Another thought: shooting animals for specimens doesn't always produce clean, non-mangled results. Perhaps in this case the gun rendered the nighthawk skin unsalvageable.

Even though Wednesday was cool and rainy, Monahan and Green were joined by friends, the Landyendorfs (elsewhere in the diary spelled Lauzendorf) from Caldwell.

The weather cleared later in the week, and the Baker team grew eager to begin adding some large animals to their collection. To that end, on Thursday they set out looking for signs of deer and bear. They found evident signs of both big game animals, but apparently not in abundance.

Friday of that week, while the men were exploring their surroundings and looking for 'something new', their camp was raided by range cows, "looting" two pounds of coffee, a half bushel of oats, one shirt and a can of cooked beans. This wanton plundering of their camp pointed up the need for the Baker team to fence their encampment while away on their hunts for specimens.

Conflicts stemming from wandering livestock were not new to Long Valley. One of the first flashpoints occurred a little over a decade earlier when cattle herded into the valley were turned loose to roam free throughout the area. The resulting stripping of crops and grazing land, along with damage done to fences and other personal property sparked some rather volatile reactions from resident settlers. Ultimately, cattlemen agreed to rent or buy several parcels of land for summer pasture. In similar fashion, early flare-ups between sheepmen and cattlemen were settled (Pottenger 1972).

Soon after the establishment of forest reserves in 1905, permits to graze livestock were available for much of those Federal forest lands bordering the valley's agricultural sectors (*The Long Valley Advocate 1905*). It's not entirely clear as to whose land the Baker team camped on – state or U.S. Forest Reserve – or whose cattle roamed through their camp. Evidently the cattle had fairly free run of the area.

SATURDAY, JULY 13 The next day, the Baker team had an unexpected visitor when the local Deputy State Game Warden strolled into camp asking to see their licenses and permits. This was of no concern to the team as they possessed all the required legal papers. But to Monahan's chagrin he was unable to find the necessary documents. The warden accepted their word and good-naturedly let the matter rest, saying it was okay. Despite Monahan's relief at the warden's amiable dismissal of the matter, in his diary he describes the warden as "- an odd fellow." This is a curious comment: did Monahan mean that the Game Warden was a peculiar or odd person, or – since Monahan was a member of the International Order of Odd Fellows fraternal organization – did he simply state that the law enforcement official was likewise a member of that group, an Odd Fellow?

The Game Warden must have been policing the area fairly thoroughly at the time. The very next day (July 14) he is implicated in a somewhat convoluted flap with a local Roseberry resident about the non-possession of a fishing license. The August 1, 1907 *Long Valley Advocate* reports a mildly embarrassing run-in between a certain J.I. Brown and the game warden. Wanting to fish the lake, but unable to locate a source for acquiring a fishing license, Brown proceeded to fish anyway finding himself chagrined when apprehended by the warden. Eventually the matter was resolved with lenience and diplomacy by both the warden and local Justice of the Peace, Eloheimo.

Sometime that same Saturday – it's unclear if it was before or after the Game Warden's visit – Monahan added a young pileated woodpecker to their collection. Curiously, in recording the woodpecker in his bird journal there appears an inexplicable gap in his numbering sequence (Note the last bird entry, No. 62 on July 6). The next entry in the sequence of bird specimens jumps suddenly to No. 68.

(Birds) *7/13* *No.68* *Ceophloeus [Dryocopus] pileatus (Pileated Woodpecker) young*

Shot from a tall pine tree.

A careful scrutiny of the bird field journal entries can explain, I think, the break in the numbering sequence. The last listing on July 6 (the ruffed grouse) is No. 62. However, that "2" looks enough like a "7" to be easily mistaken (a week later) for "67". That is the best explanation I can offer for this puzzling number omission.

SUNDAY, JULY 14 It wasn't until the next day that Monahan finally located his collecting permit and prepared a copy to be mailed to the game warden in Cambridge. Anticipating the next day's trip to the post office, the men spent part of their quiet Sunday writing letters.

That ends the diary records for the day. It was Sunday, typically a quiet day in camp. However, Monahan's field journal records the following on this date, indicating the collection of a female Swainson's thrush plus her set of eggs:

~~~~~~~~~~~~~~~~~~~~~

(Birds) *7/14 No. 69 Hylocichla ustulata swainsonii [Catharus ustulatus] (Russet-backed Thrush)* [Swainson's thrush]

Taken from waters edge Payette Lake Ida. On the nest, 3 eggs unincubated. Nest 3 ft from water in Mtn Willow bush

7/14 No. 93 Buteo [jamaicensis] harlani (Harlan's Hawk)

(Mammals) *7/14 No. 13 (Jumping or Kangaroo Mouse)*

Taken in a meadow on Payette Lake Boise Co. Ida Very shy & rare

~~~~~~~~~~~~~~~~~~~~~

This mammal is most certainly *Zapus princeps* the western jumping mouse. This long-tailed little rodent does somewhat resemble the unrelated desert-dwelling kangaroo mouse, but is strictly a mountain species. As Monahan's first specimen of the kind, his name-listing reveals a little uncertainty in its identification.

The Harlan's hawk specimen requires some clarification. First, I need to point out that raptors have long been one of the most challenging groups of birds for species recognition. The Harlan's is a very dark-colored hawk in the *buteo* (broad-winged, broad-tailed) group and most closely resembles the red-tailed hawk. Despite its similarities to the red-tailed, it was considered for many decades a separate species, as it was in 1907. Because red-tailed hawks show a high incidence of "melanism", the dark color form, those darker individuals look confusingly similar to the Harlan's hawk. Harlan's hawks do stray into Idaho during winter months but in summer during the time that Monahan records this collection their breeding population is restricted to Alaska and northern British Columbia. I think it is safe to say that this hawk (No. 93 above) was a melanistic

(dark morph) red-tailed hawk. It would be an easy mistake to make. Of course today, the Harlan's has been lumped with the red-tailed hawk as a northern subspecies.

MONDAY, JULY 15 Anticipating further word from Will's brother Ray about his up-coming visit, the team went into town that morning to check for mail. There was no word from Ray, but while in town they picked up some needed provisions – bread, butter, cheese and catsup – plus the all-important nails for constructing their camp fence.

Returning to camp they busied themselves in putting up a good perimeter fence to safeguard their campsite from further damage from the free-ranging cattle of the area. They had suffered once from their bovine marauders, and wanted no part of any return visits.

Later that day, the men set out hunting for specimens. Growing more familiar with their lakeshore surroundings they were discovering more and more signs of wildlife. The diary records that the men found tracks of deer and lynx, and took a set of eggs from a "Russet-backed" (Swainson's) thrush nest. They even chanced upon one of their long sought-after species, mallard ducks. The Baker team had been on the lookout for these waterfowl since first entering Long Valley. However, despite their good luck at finding the species, they were not so lucky in actually collecting one. They shot down a young mallard but their lack of a boat prevented its retrieval. Nevertheless, the wildlife sightings of the day buoyed their hopes and they laid plans for an even longer trip the following day.

Note the discrepancy in dates on taking the Swianson's thrush, listed in the field notes as collected the previous day. This begins a series of one or two day discrepancies between what is recorded in the diary and the field journal collection notes.

> (Birds) *7/15 No. 70 Merganser [Mergus] americanus Juv (Am.*[Common] *Merganser)*
> *Taken from flock of 7 on Payette Lake – (Iris white)*

TUESDAY, JULY 16 Awakened the next morning by whipping sounds on their tent, the team arose to find their friend Landyendorf outside delivering a morning wake-up call. Since Monahan and Green were planning another big day hunting and exploring their way to the upper end of the lake, their friend had arrived to join them, perhaps offering welcome guidance in some new areas.

While preparing breakfast Monahan was miffed to discover that a chipmunk had entered their tent and eaten his newly collected thrush eggs. Surely that was an ill-advised act to perpetrate on a camp equipped and intent on collecting critters. I'm guessing that the guilty rodent and many of his next-of-kin were doomed to specimen cases over the following days.

The three men ate breakfast, then loaded their gear into a boat and set out for the upper end of the lake. By noon, they arrived at Fall Creek, where they sat on a log and ate lunch. In the course of their travels Will tried his hand at fishing. The younger member of the Baker Team must have finally caught on to Long Valley angling, for unlike earlier unsuccessful attempts, on this afternoon he caught six nice trout.

A passing afternoon shower doused the men, but the weather quickly cleared and around 3:00 P.M. they were headed back toward camp. It is on this return trip that the team reports what is to me one of their most notable wildlife encounters. And I'm not referring to the nice young merganser specimen that Monahan reports collecting. Rather, it was the prize that narrowly escaped them. As Monahan writes: "Saw a Great Grey Owl – gun jammed on me and it got away – fine specimen."

This event is almost symbolic of the great gray's near-mystical elusiveness. Scattered throughout the upper reaches of North America, the great gray's rarity and reclusive habits have long brought it legendary acclaim. We know little about the owl's connection to its very first observers, the Native Americans. I'm guessing it was noble and reverential. At one of the few modern Native American pow-wows that I've attended, I clearly recognized several great gray tail feathers decorating one participant's headdress. Like the highly prized white-blotched eagle feathers, I expect large owl feathers have long been a part of North American Indian cultural attire. Our country's earliest stories of this owl have emerged from keen-eyed trappers and backwoodsmen who long ago hailed the bird as the "Phantom of the Northern Forest", and "Gray Ghost of the Woods".

The great gray's connection to Idaho runs as far back as our state's earliest scientific footprint. Their presence in Idaho was first reported a century before the Baker trek, when the Lewis and Clark Expedition collected a specimen at Kamiah, along the Clearwater River on May 23, 1806 (Burleigh 1972). Monahan's notation in this 1907 diary stands as only the second report of the owl in Idaho. This gray ghost has surely skulked Idaho's wooded meadows since time immemorial, yet this 1907 gun-jamming day in Long Valley provided only its second reporting by a scientist in Idaho.

Today, a century after Monahan's brush with the species, the great gray owl's Idaho status remains about as nebulous as it was then. As detailed in my 2008 book, *Dead Owls Flying,* Long Valley formerly hosted respectable numbers of great grays. In our initial studies of the owl in 1986-88, my students and I accounted for at least 30 great grays in two small surveyed sectors of the valley. A decade later, our follow-up study found those numbers, amid increasingly fragmented habitat, cut in half. Sadly, as I write this book, the prolonged and unbridled development of those same portions of the valley have now decimated the numbers of great grays. The occasional photo and isolated reports of the owl currently trickling out of Long Valley are restricted to fewer and fewer sites. Ironically, a couple of those sites of lingering great gray presence are scant miles from Monahan's 1907 sighting.

Following their great gray encounter that afternoon, the men returned to camp, weary from the day's outing. Their excursion had been largely a bust. They had seen no deer tracks or bear signs, had missed the opportunity to collect a great gray owl, and were rained on twice. Although unrecorded in the diary, somewhere along the way they did successfully take a new bird specimen – the only one of its kind for the summer expedition. In fact, the following specimen record brings to light several fascinating details. First, it comprises one of the few lengthy discourses found within Monahan's field journals (second only to the shooting down of the oriole nest from a tree at Robie Creek). As already evident across the previous pages of this book, Monahan's field journal notes are typically very brief, methodical notations. But this recording is an exception.

> (Birds) *7/16 No. 71 Dryobates [Picoides] villosus (Hairy Woodpecker)?*
>
> *Taken in deep pine woods on Payette Lake. I call this D. villosus on my own analysis, however Coues says the pure D. villosus is not found west of the Rocky Mts; the D. v. harrisi taking its place. Reed says "D. v. harrisi is found from California & B. Columbia to the Pacific Coast." I am quite sure that this specimen (No. 71) must be the Eastern type*

Note the unusual presence of a question mark behind the

common name – one of only two question marks appearing in his field notes. Here, Monahan's uncertainty isn't a question of whether the bird was indeed a hairy woodpecker, but rather, which of the several subspecies it represented. The D. v. harrisi race (Columbian hairy woodpecker) was described in Elliott Coues (1903) as "smoky bellied" instead of white on the under parts. Monahan must have noticed the lighter colored under parts on his specimen which was highly reminiscent of his Eastern hairy woodpecker. Today's ornithologists recognize Idaho's most likely occurring race of hairy woodpecker as the lighter-colored "interior west race" (National Audubon Society 2001).

As inconsequential as Monahan's quandary over woodpecker subspecies seems, it inadvertently serves to announce our first clues to at least two of Monahan's bird references he must have carried with him. In addition to his reference to Coues, note he also cites

Reed. That would certainly have been Chester A. Reed, whose two books in 1906 were (1.) *Bird Guide: Water Birds, Game Birds and Birds of Prey East of the Rockies,* and (2.) *Bird Guide: Land Birds East of the Rockies* (published by Doubleday, of New York). Thus, from the several possibilities discussed in the Preface, we finally discover some of Monahan's sources.

By bedtime, after drying-out by the fire, eating a mess of freshly caught trout, and having a good smoke, Monahan was feeling much better.

WEDNESDAY, JULY 17 The men rose at 6:30 that morning, planning to scout out good places along the shoreline to set some lynx traps. After exploring the area for awhile and finding no fresh signs, the men finally opted not to set traps. The foray was not a total failure, however, as Monahan discovered another nest of thrush eggs, to replace his earlier set that had been predated by the camp chipmunks.

That afternoon the men visited the Post Office still hoping for word on the anticipated arrival of Will's brother. No luck with the mail but while in town they picked up some fresh bread and milk.

On the heels of yesterday's rain showers, Monahan notes in the diary that the day's weather was cool enough for October.

> *(Eggs) 7/17 No. 17 Hylocichla ustulata swainsonii [Catharus ustulatus] [Swainson's thrush]*
> *Number in set 3; situated low bush on lake shore; common here; nest - strips of bark & roots. Boise Co. Idaho Lake Payette.*

THURSDAY, JULY 18 Monahan and Green had resided nearly two weeks at their present campsite, and they were growing increasingly anxious to take some big game specimens. To that end, the following day Monahan and Green headed back up into the high country in search of deer and bear. In the course of the day they trekked high enough to find themselves in patches of snow. They did see deer tracks, but still found no signs of bear. On the way back down to camp the two men somehow became separated, returning along different routes. Monahan reached camp first. By the time Will and Jo, Monahan's dog, straggled into camp, Monahan had caught a nice mess of trout for supper. The three – both men and their dog – wearily called it a day, disappointed at having heard nothing from their anticipated visitor, Ray Green. As soon as they heard, they planned to head for Meadows to meet him at the end of the rail line.

According to the diary account, the specimens listed immediately below were collected the next day, however, I record them here as they are dated in Monahan's field journals. Which source is correctly dated I have no way of knowing.

~~~~~~~~~~~~~~~~~~~~~

> (Birds) *7/18 No. 72 Bubo virginiana (Great Horned Owl) Juv*
> *Taken in deep woods on Payette Lake Boise Co. Ida*

> *7/18 No. 73 Parisoreus [Perisoreus]* canadensis *(Canada* [Gray] *Jay)*
~~~~~~~~~~~~~~~~~~~~~

Taken in deep woods near Payette Lake Boise Co. Ida – a rare bird in this county In this season.

7/18 No. 74 Sphyropicus [Sphyrapicus] thyroideus (Red-throated Woodpecker)

[Williamson's vsapsucker] *Juv*

Taken from fir tree on Lake Payette Boise Co. Ida

~~~~~~~~~~~~~~~~~~~~~

FRIDAY, JULY 19 The Baker team left immediately after breakfast on one of two sorties to find some owls that had been reported to them. Which species of owl they were after is not stated, but they found none on their first trip. On that first outing, they were successful however, in taking a woodpecker specimen. Again, the diary does not specify which species of woodpecker they collected.

Soon after returning to camp, the Baker team's Dutch friend, Landyendorf, showed up. He and Monahan again went searching for the owls. This time they were successful in locating and collecting a great horned owl, judging from the field journal records (dated the previous day, however).

As the day continues, so do the number of mystery birds. After lunch, the three men head for the upper end of the lake. It's not clear if they went by water or by land, but in either case they spent part of their time in the woods attempting to climb a tree to reach a "red head's" nest. Recall that this "red head" bird species is one of the continuing bird mysteries running through the diary. Was it a redhead duck that they were after? Probably not, if its nest was located in a tree. By all accounts the redhead duck is an inveterate ground nester. Before my "eleventh hour" acquisition of Monahan's field journals, I could only wonder: was this "red head" a reference to the male western tanager; the yellow, sparrow-sized bird bearing a reddish head? Despite their enterprising construction of a make-shift ladder – leaning a smaller tree against the nest tree and
~~~~~~~~~~~~~~~~~~~~~

strapping them together – upon inspection, the "red head's" nest was empty.

The final mystery bird collected that same day is recorded in the diary simply as a "*...female red throated*". Again, the precise identity is not given and therefore open to speculation. And here again for 36 years prior to my acquiring Monahan's long lost field journals, I had to puzzle over this mystery. Were they referring to a red-throated loon? Not likely, as this coastal bird is rarely reported in Idaho. One of the few surviving 1907 Monahan specimens that I salvaged from Baker University back in 1976 was a male Williamson's Sapsucker, which bears a characteristic, small but distinctly red throat patch. Could this have been the "red throat"? Not unless they misidentified it as "female", which bears no such marking. By the way, the distinctly different coloration of the genders (the male is mostly black with a red throat patch, the female is mostly brown and barred on the backside) of this species fooled earlier naturalists from the very beginning. In 1851 John Cassin, discovered a brown ladder-backed colored female sapsucker and named it Cassin's Sapsucker. Six years later another pioneer naturalist, John Strong Newberry collected a black male sapsucker, and seeing little resemblance to any other sapsuckers, he considered it a new species. He named this "new" species for his Survey leader, army lieutenant Robert Stockton Williamson (Evens 1993). It would be 20 years later before more careful observations proved these two very dimorphic sapsuckers to actually be different genders of the same species.

Fortunately, the above diary listing is corroborated and clarified by the field journal notation below. In 1907 the Williamson's sapsucker still went by the name, "red-throated woodpecker"

There remains one other notable event recorded that day; they missed shots at a "fish hawk" (osprey). As common as this raptor is around the area's lakes and water courses today – despite its near extermination during the DDT binge of the 1950's and 60's – it's a little surprising to find this is one of only two reports of the species throughout the entire diary. Perhaps, this was simply the only notable opportunity to collect one, even though they were observed on other occasions.

Several chipmunks were pronounced dead that day. Since the mammal collection journal notes do not record any small chipmunks taken until later into September, I suspect this was serving retribution to these small bandits around camp for their pilfering of Monahan's thrush eggs three days earlier.

After several previous days of intermittent showers and cold weather, this day ended warm; inviting the dreaded accompaniment of swarms of mosquitoes.

> (Birds) *7/19* *No. 75* *Sphyropicus [Sphyrapicus] thyroideus* [Williamson's Sapsucker] *Juv*
> *Notes as in # 74*

 SATURDAY, JULY 20 & SUNDAY, JULY 21 Over the weekend the men stayed close to camp, awaiting word from Will's brother Ray. On Saturday, Will went into town, checked the mail and bought some molasses, bread and butter. Although Monahan scored with three letters that day, Will received none. Still, there was no word from Ray.

It seems almost incredible that as Monahan awaits Ray Green's step onto their Baker Expedition horse-drawn 'Science Wagon', a mere 62 years later on this date – only six years after Monahan's death – American Astronauts would step onto the moon. How foreign would this thought have been to the two members of this pioneer expedition?

Nothing is said in the diary of collecting specimens that weekend, yet the field journals record the following:

~~~~~~~~~~~~~~~~~~~~~~

> (Birds) *7/20* *No. 20* *Sphyropicus varius nuchalis [Sphyrapicus nuchalis] (Red-naped Sapsucker)*
> *Taken on Lake Payette Boise Co. Ida*
~~~~~~~~~~~~~~~~~~~~~~

7/20 No. 94 Podiceps auritus californica [Podiceps nigricollis] Am. Eared Grebe [eared grebe]

(Eggs) *7/20 No. 18 [Dendroica coronata?] Myrtle Warbler* [Yellow-rumped Warbler?] *Taken from Fir tree near lake side*

~~~~~~~~~~~~~~~~~~~~~

I'll discuss this highly unusual if not questionable egg collection below where it appears in the diary account.

On Sunday the team spent part of the day fishing at Fall Creek where they caught enough fish for that evening's supper and breakfast the next morning.

MONDAY, JULY 22 This day breaks warm and calm, and as Monahan describes, "...lake smooth as glass." In fact the weather is mild enough that the mosquitoes are becoming an increasingly vexing problem, forcing the men to build deterring smudges both morning and night. Yet, despite the torment of the mosquitoes, the men seem in fine fettle, as Monahan notes: "Feeling fine, eating like hogs on full feed." The men remain in somewhat of a holding pattern, hoping the next day will bring them word from Ray Green.

Three notations in the day's diary entry intrigue me. First, Monahan records that he spent much of the day "monkeying" with a camera. Since the Baker team carried a camera on their expedition, there is the tantalizing possibility that he and Green may have recorded some aspects of their expedition on film. This, despite the fact there are no such photos accompanying the published diary. In all of Professor Sokolofsky's research on the expedition, preparatory to publishing the diary, he apparently found no photographic record of the expedition. Nor did he produce any Baker University photos of Monahan and Green from the time period of the expedition. The few pictures used to illustrate the published diary are simply non-related photos depicting similar subjects. Later in the summer,
~~~~~~~~~~~~~~~~~~~~~

the diary also records failed attempts at photography. In all of my research on Monahan, including his son's search of the family archives, no photos from the 1907 expedition have ever been found.

The diary also records a couple of remarkable specimens collected that day. One was an "eel". This unusual fish would have been the Pacific lamprey (*Entosphenos tridentatus*). Though unlovely to look at and derided by most fishermen, the lamprey is the darling of comparative anatomists. It flaunts an intriguing combo of the ancient and modern in one perfected form. It's a truly modern, highly specialized parasitic fish, with a round sucking mouth for attaching to the sides of large host fish. Then it plies its rasping tongue to open a wound from which it feeds on fluids from its victim. Moreover, the lamprey showcases a number of primitive features – no jaws, unpaired fins, rudimentary skeleton and simple internal organs. Once native to the major water courses of the Pacific Northwest and migrating with the salmon, the lamprey has been absent from the Long Valley area since the construction of the first dams on the Columbia and Snake Rivers. Because of its declining numbers, due to the difficulty of navigating the numerous Columbia Basin dams, this lamprey is currently a candidate for federal listing as a threatened and endangered species. Probably few people lament or even notice the loss of this particular anadromous fish to our present day gauntlet of hydroelectric dams. Nevertheless, this 1907 report reminds us of one of Idaho's diminished life forms. What was once part of the whole is now missing, and begs the question: what did we lose with it?

The other intriguing collection of the day was the nest and eggs of a "Myrtle warbler". This small, multicolored bird is nearly identical to the Audubon form of the yellow-rumped warbler. The whiter wing patch and yellow throat of the Audubon contrasts with the white throat and scant two wing bars on the Myrtle. Formerly, these two warblers were deemed separate species, but today with advanced DNA analysis, they are judged to be merely different color forms of the single species, the yellow-rumped warbler.

No matter their taxonomy (scientific grouping), the Myrtle form typically nests only in the northern and eastern U.S.A.; and though the Myrtle migrates through Idaho, as far as I know, there are no nesting records within the state; this casts some doubt on Monahan's identity of this warbler nest and eggs.

In the 1907 bird literature sources, such as Coues' 1903 books, the two warblers are clearly differentiated so I have no good explanation for this unusual nesting record, except to embrace it as valid. One could argue that while collecting the eggs, Monahan likely saw only the female and if it were the Audubon form, her normally paler yellow throat (less yellow than the male) may have deceived him as appearing white.

TUESDAY, JULY 23 The Baker team finally received word that Ray Green was on his way to join them. He was coming by train, and since the railway system had not yet made its way into Long Valley (that wouldn't happen until 1912), it meant that Monahan and Will Green would need to meet him at the rail terminus in Evergreen, south of Meadows. Ray was planning to make connections with them on the following Monday, July 29.

There was no need for the Baker team to depart for Evergreen until the following day, so the men spent part of the day hunting for evening grosbeaks. Though they found no grosbeaks, they did discover three nests of western wood pewees. Monahan records his disappointment that all three nests contained nestlings instead of eggs, which was the real prize they were after. Because it was late in the summer, nearly August, it is little surprise that the egg-collecting period was rapidly waning. Most of the birds in the area were well through their nesting season. In fact, Monahan's journals record no further egg-collecting after taking the Myrtle warbler eggs. In all, he had taken 18 sets of eggs from 16 bird species thus far in this Idaho expedition.

This aspect of the expedition's specimen search provides us a reminder of what was then still a very common practice, yet one totally foreign to us today: the collection and study of bird eggs, or oology. Certainly, it had its rightful place in museum collections; such was the case with Monahan's repeated interest in locating nests – complete with eggs – of various Idaho birds. But even before 1907 widespread egg-collecting had expanded well beyond museum and scientific collections. It raged unchecked in many parts of the civilized world. For the most part, they were trendy private collections, akin to those of bird skins, butterflies and sea shells. Aristocratic households boasted ornate cabinets full of colorful egg collections. It was in vogue across Europe and America. However, this universally popular hobby wasn't necessarily restricted to the wealthy. What curious youngster or adult couldn't go out and locate a neighborhood bird nest and collect its eggs? On a small scale, it wasn't a prohibitively expensive hobby. Yet, once egg collections became fashionable and quite prestigious there emerged an enterprising market. Egg collectors soon abounded and operated on the frontier premise that bird eggs, like all else in nature, were an inexhaustible resource. Not surprising, the eggs of rarer birds became quite valuable on the market. Scott Weidensaul (2007) notes that early, trendy magazines like *Ornithologist and Oologist* openly advertised this egg market, wherein a blown egg of a California Condor might fetch $350, the equivalent price of a small house. Somewhat symptomatic of the excesses of this perversion, one oologist boasted of owning 180 sets of peregrine falcon eggs, totaling more than 700 eggs.

As obscenely profligate as those private egg collections may seem to us today, they were small potatoes compared to some instances of mass marketing of eggs. For example, in 1901 about 630,000 eggs of the jackass penguin were collected on South Africa islands. Even as late as 1946, in the Seychelles Islands (Indian Ocean) 1, 996,400 tern eggs were collected (Welty 1988). Certainly, the little bit of egg-collecting recorded by the Baker team was a far cry from the dark side of oology. It was simply a scant exercise of scientific documentation.

·9·

TO THE MEADOWS TO MEET THE TRAIN

(JULY 24–JULY 31)

WEDNESDAY, JULY 24 At breakfast time the next day, their friend and neighbor rancher Walter Gaekle, rode into camp and reported seeing a bald eagle perched in a tall pine not far away. Apparently, several earlier attempts by Monahan to collect this bird had failed. At this fresh opportunity, the men grabbed their guns and headed out. In a matter of minutes a single shot from their trusty 40-82 caliber rifle brought the bird crashing down. Monahan was elated. He describes his prize collection in his diary that day:

"He is a handsome bird measuring 6 ft., 10 inches in extent. He is a two year old –white as snow on head & neck & tail. This is one of the much-sought for things by me for the Baker museum. His eyes are white & his bill a beautiful naples yellow – body brown. This specimen is worth $10.00 to us at least."

Monahan's plumage description and assessment of the eagle's adult status are clear and accurate. Today, however, we better understand the molting process of this species and now know that bald eagles don't attain their adult white head and tail plumage until four years of age, instead of two years, as Monahan stated. At the time of Sokolofsky's research for publishing Monahan's Diary (1969) this bald eagle was still identifiably on display in the Baker Museum. Through the years in my many readings and study of the published diary, I wondered if that bald eagle yet remained on the Baker campus. Finally, as I describe above in Chapter 5, to satisfy that curiosity, in August 2009 I contacted Baker University biology professor, Dr. Roger Boyd, and posed that question. You know the rest of the story: that simple inquiry dramatically altered the course of history on this book, enriching its content exponentially.

Evidence strongly suggests that the bald eagle Monahan collected that day still resides at Baker University, though it no longer bears any tag or reference to 1907 or Idaho. A fascinating line of evidence relates to the uniquely peculiar stance and position in which the old eagle is mounted. It perfectly duplicates that of the bald eagle illustration pictured in the 1903 Elliot Coues bird book that Monahan used (see photos). Dr. Roger Boyd, Ornithologist, taxidermist and curator at Baker University shares my contention that in all probability Monahan shot that eagle specimen and mounted it using his old Coues book illustration as a model.

Following the morning excitement, while Monahan jubilantly skinned out their eagle prize, Will Green loaded the wagon and prepared for their departure to The Meadows. Some of their camping gear had already been transferred to their neighboring rancher friends, the Gaekles, for temporary storage.

By 9:00 A.M. the Baker team was on the road. The Meadows valley is approximately 11 miles northwest of Lardo, mostly downhill, dropping some 1200 feet in elevation. The road to the lower valley was in fine shape and the men made good time, stopping for lunch just outside Meadows. When they rolled into the town, Monahan inquired about a Mr. Meyers with whom he had made prior arrangements to obtain a beaver specimen. Unsuccessful in locating their beaver man, the team journeyed on and found a suitable camping spot at the head of the Little Salmon River.

In his diary, Monahan describes the Meadows as a basin four by twenty-five miles long, settled by intelligent, industrious folks from Kansas, Nebraska, and Missouri. He records the crops as primarily timothy and alfalfa hay, oats and barley – partly irrigated and partly rain supported.

The lone specimen recorded for the day in Monahan's field journals is the bald eagle he collected that morning before leaving their Payette Lake camp. This is one of the few specimens for which he makes no determination of gender. Instead, he records various body measurements for later use in preparing the taxidermy mount of this prize bird.

> (Birds) *7/24 No. 79 Haliaetus [Haliaeetus] leucocephalus Bald Eagle*
>
> *l of body 9.5 l over breast from neck to tail 15 l of breast from shoulder to shoulder 13 l of neck 5 width of girth around wings 17*

THURSDAY, JULY 25 Breaking camp that morning, the men apparently anticipated doing some fishing along the way because before leaving the grassy meadows they made a point of gathering some grasshoppers for bait.

Journeying on toward Evergreen, Mud creek must have looked irresistibly enticing, because it wasn't long before they stopped to fish. The men split up and fished different directions along the creek – Monahan followed the stream downward and Will fished upstream. There's no mention of which man fared better but they did well with a combined catch of a dozen fish. Taking a respite from their angling, the two men returned to their wagon and drove a mile up the west fork of the creek where they found on old corral in which to camp.

After setting up camp, they feasted on their freshly caught trout, and soon headed up creek from camp for more fishing. There they fared even better, and the diary reports a catch of 34 fish, some a foot long. After cleaning and salting their afternoon catch, the men groomed and hobbled their horses. Though the valley seemed to be nearly overrun with cattle and horses, the men were quite happy with their selected camp spot. In addition to the amenities of fish, good water, grass, and shade, perhaps the camp's most inviting element was the relative absence of mosquitoes. Even the brief afternoon sprinkling of rain likely offered welcomed relief from the heat, and no doubt settled the dust as well.

On a brief historical note: stagecoach lines ran up through many of the major settlements and mining areas of early Idaho. Monahan mentions passing a stage driver flirting with girls that day. Since Ray Green could have ridden a stage from Evergreen to Lardo, I suppose the fact he didn't, and instead, was met by the Baker team at Evergreen, was a matter of shaving traveling costs for their visitor. Relative to this might be the fact that in 1907 our country was going through one of the greatest financial downturns in our country's history – probably surpassed only by the Great Depression and our current (2008-10) economic malaise.

Whatever the motivation for this lengthy side trip, one must recognize that their rendezvousing junket to Evergreen certainly provided an opportunity for the expedition to sample specimens from a very different mountain locality, lower in elevation and some distance from their regular lakeshore camp.

FRIDAY, JULY 26 The day dawned with light rain showers, so the men slept late. Before long, the cessation of rain and promise of exploring new landscapes rousted the men out for their morning camp duties. While Green started breakfast preparations, Monahan went after the horses. Being unfamiliar with the terrain, the older man wandered far and wide before finally locating the horses in a small ravine back near their camp. His extended round-up delayed breakfast but no doubt rendered Green's corncakes, trout and oatmeal all the more tasty.

It was 11:30 by the time breakfast was over and the men headed back up creek for more fishing and hunting for new specimens. It was a productive three mile excursion. Green caught some trout and along the way the men collected a sharp-shinned hawk and a "red head's nest". Again, I assume the identity of this red head to be the western tanager. On the way back to camp the team hunted the hillsides for grouse but found none.

However, after returning to camp the men collected a couple of woodpeckers, one of which was perhaps a prize specimen of the summer. Along with the rather common northern flicker they collected a rare three-toed woodpecker. Neither the diary nor the field journal entries give specifics as to which species of this rare woodpecker was collected - the American three-toed or black-backed woodpecker. Formerly, both bore "three-toed" in their names and both are uncommon birds in the central Idaho mountains. The fact that Monahan leaves blank the space in which he customarily wrote the scientific name of his specimen – entering only "3-toed woodpecker" – suggests his uncertainty about which of the two closely related species he had.

As the afternoon got hot, the men busied themselves around the shaded camp. While Green prepared a late afternoon meal, Monahan tended to the horses and turned them out to graze. After dinner Monahan skinned out the woodpeckers and the two men relaxed and smoked. In vicinity of camp they observed a pygmy nuthatch, but make no mention of collecting it.

~~~~~~~~~~~~~~~~~~~~~

(Birds) 7/26 No. 80 Accipiter velox [striatus] (Sharp-shinned Hawk)
Taken in Boise Co. Idaho rare here

7/26 No. 81 3 toed Woodpecker
Taken in Boise Co. Idaho

7/26 No. 82 Colaptes cafer juv (Red-shafted [Northern] *W.pecker* [Flicker]
Boise Co. Idaho

~~~~~~~~~~~~~~~~~~~~~

It is interesting to note that despite the diary account of collecting a red head's nest that day, there is no record of such in the field journals. At this late date it is possible that the nest may have contained hatchlings instead of eggs.

Diary notations for that day suggest that it was also during this down time around camp that the Baker team devoted some time to collecting insects. This is only the second mention of insect collecting in the entire diary. Even so, from the newspaper article at their early summer departure from Boise plus previous diary references to some of their field gear including insect nets, we know the team had some entomological interest and may have done more insect collecting than they record in the diary. The field journals record nothing of invertebrates of any kind.

SATURDAY, JULY 27 The hot weather they encountered in the lower elevation of the Meadows area may possibly have spurred the notice of insects. The men spent a hot, quiet day in camp, complaining of both the heat and the hordes of flies – especially the blue-bottle flies. Recall the men's immediate relief of escaping the scourge of hungry mosquitoes after entering the meadows valley. There seemed now to be a trade-off of pestilence: flies for mosquitoes. It would be several decades before DDT would wield its deadly blow to insects and blanket its insidious pall over the environment. In 1907, flies ruled.

Green took off on a fishing excursion while Monahan did some follow-up work on the prized bald eagle skin they had acquired prior to leaving the lakes. I can imagine the pesky flies vying for an opportunity at the eagle skin, even as Monahan was likely applying a preservative to the skin that would forestall any such insect invasions or decomposing microbes.

Green's fishing foray was highly successful, returning to camp with a whopping catch of 19 nice trout. As Monahan and Green dined on the trout for supper, their thoughts turned toward home and their Kansas friends, wishing they could be there to enjoy the feast with them.

The Baker team planned to journey on to Evergreen the next day where they would meet Ray Green on Monday. As was their periodic practice, they ran the "Ol' Science Wagon" into the creek to soak the wheel spokes for smoother, tighter running. Perhaps to beat the valley heat and spruce up for a trip into the town of Evergreen, the men indulged themselves with a bath in the clear, cold stream. No doubt both the wagon and the wagon-drivers fared better for their cool soaking.

At the end of the day Monahan signs off the diary with a lighthearted note about going to bed, "...to dream of killing bears, deer, cougar, & Mt. goat." I expect that for these two Kansas naturalists, over a thousand miles from home, collecting wild and wonderfully new creatures in a remote and rugged landscape, such thoughts danced across many moonlit nights before they fell asleep. I imagine too, that some of their dreams highlighted wonderful discoveries and collections for the Baker Museum. They were, after all, one of the last of the West's pioneer biological expeditions. Theirs would be Baker's University's last big expedition.

SUNDAY, JULY 28 The Baker team trekked on to Evergreen, deviating from their previous policy of not traveling on the Sabbath. This was only the second time they elected to travel on Sunday. In this case, I suspect they were eagerly anticipating the reunion with visiting guest, Ray Green, and were anxious to reach Evergreen and get settled somewhere on the town's outskirts. Because the railroad line ended there, the small Evergreen community was the dropping off point for goods, mail and people traveling into the area. By Monahan's description, in 1907 the town consisted of little more than a store, post office and train station.

The weather remained hot, with a threat of afternoon showers. For a short while the two men fished the Weiser River, but were so disappointed in the seven small trout they caught that they disgustedly tossed them back into the water.

Their sour fishing episode was quickly sweetened when they noticed some of the locals out picking huckleberries and service berries. Monahan soon learned from one of them that, in this high country, the people depended on these wild plants for fruit. And apparently that summer of 1907 was a good fruit year, cycling about every four years. Monahan and Green enjoyed the wild berries so much they determined to go berry picking the next morning. That would help them pass the time since they didn't need to pick up Ray Green at the train station until around 1:15 in the afternoon.

There is virtually no botanical information recorded in Monahan's diary or field journals; which is lamentable. Idaho mountains are a-blush with wildflowers all summer long. With this rare allusion to plants of any kind – huckleberries and service berries – it seems appropriate here to insert at least a snapshot of wildflowers most likely sprinkling the Evergreen/Meadows countryside.

At this time of the summer you'll see tall clumps of white, lacey hillside ocean spray and competing heads of elderberry flowers standing tall along forest edges. Open grassy areas are spiked with tall corn-like stalks of California false hellebore. This robust lily's showy clusters of white flowers belie the poisonous danger lurking below in its rootstocks. Livestock beware! Bright yellow sprays of goldenrod nod shoulder to shoulder with rosy spikes of pink spirea. By now summer has spread the meadows and open forest floor with blue lupine – "blue bonnets" to the eyes of the pioneers. Hinting of summer's creep toward autumn, rusty stalks of sour dock or Indian tobacco are already ripe in angular fruit. A century ago, would common (wooly) mullein, Canada thistle, yellow salsify and common yarrow have stood forth in such abundance as they do today? The Baker team collected plants along the way. Presumably lost, their record would have provided us a flowery view into Idaho's botanical past. I lament that loss of fragile beauty.

Puny trout and a tasting of wild berries were apparently the only take of that day; and those fish had been scornfully tossed back. They had exhausted their supplies of butter, bread and coffee, but would wait until the next day to visit the Evergreen store while meeting their friend at the train.

Elmer Perry Monahan, Senior photo—1905

Courtesy of Baker University Archives.

Elmer Perry Monahan Jr. holding a grandchild, ca. 2004

The Baker Museum in its prime.

With the closing of the Baker Museum, most specimens in the collection were put up for auction. Here, a proud couple displays a bison they purchased. (The bison was collected in 1912 in Yellowstone by a separate expedition.)

Courtesy of Baker University Archives.

Courtesy of Baker University Archives.

Courtesy of Baker University Archives.

The Baker Museum in its dilapidated condition just prior to its demolition around 1976, the same time that the author acquired the five remaining 1907 Idaho specimens.

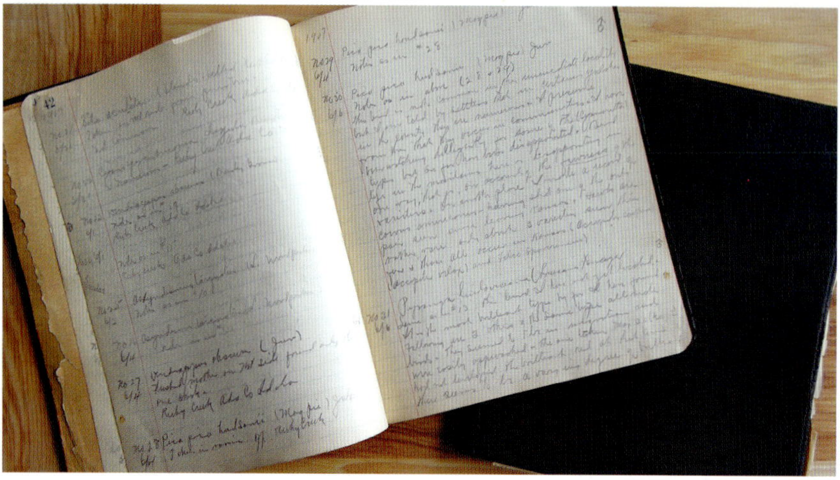

The two old field journals discovered among dusty boxes of plant specimens from the old museum. They lay buried for over a century and when recently discovered, no one knew what they represented. Then astoundingly, an inquiry from the author to Baker University biologist Dr. Roger Boyd, directed attention back to these old forgotten journals that had been lying around the department. They turned out to be Monahan's field notes from the 1907 expedition – totally different from the diary that was found decades ago and published elsewhere in 1971.

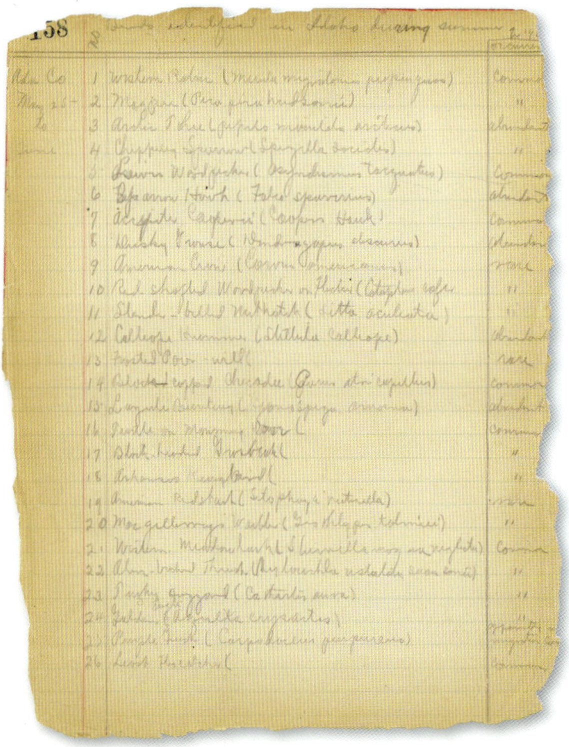

A loose page from Monahan's field journals, titled: "Birds identified in Idaho during the summer of 1907". Inscribed in the upper left hand margin is: "Ada Co. May 25 to June..." It appears to be a list of bird species they began to compile while camped at Robie Creek, May 25 – June 22, and then discontinued once they headed north to Payette Lake.

Lewis Woodpecker (#10) collected on May 29, 1907 and a Western Tanager (#34) collected on June 6, 1907 at Robie Creek. These are two of only six specimens remaining from the 1907 expedition. Journal entries of this collection can be seen on the page from Monahan's old field journal.

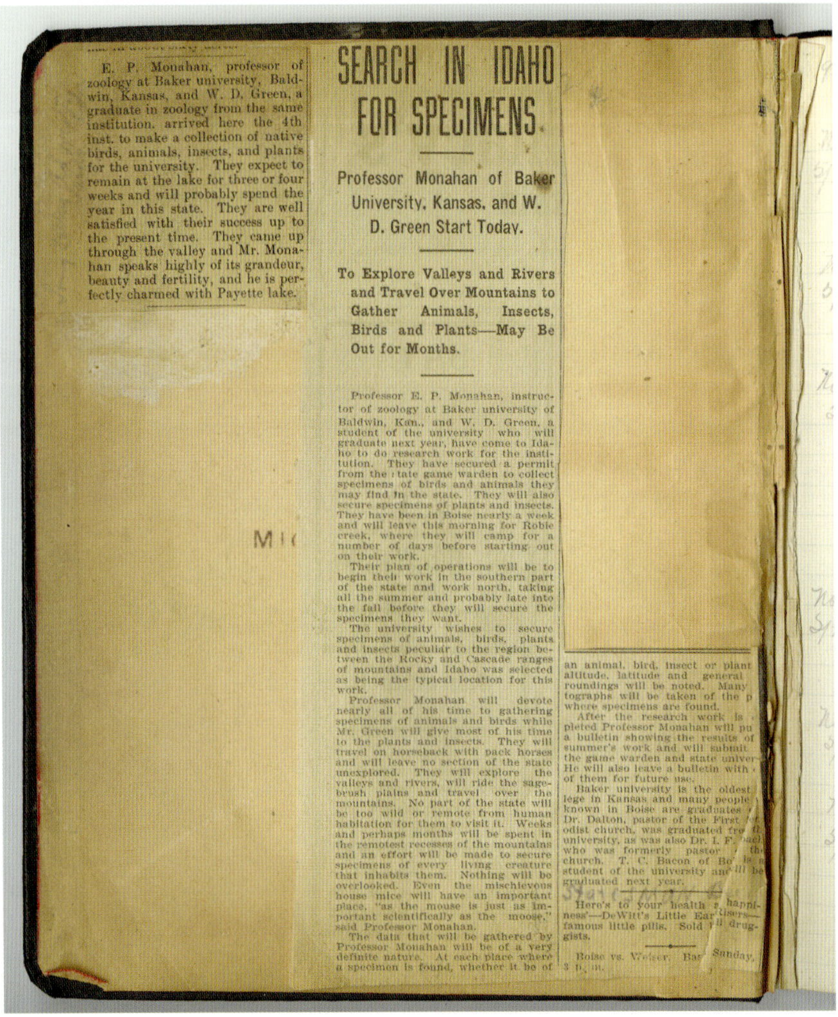

E. P. Monahan, professor of zoology at Baker university, Baldwin, Kansas, and W. D. Green, a graduate in zoology from the same institution, arrived here the 4th inst. to make a collection of native birds, animals, insects, and plants for the university. They expect to remain at the lake for three or four weeks and will probably spend the year in this state. They are well satisfied with their success up to the present time. They came up through the valley and Mr. Monahan speaks highly of its grandeur, beauty and fertility, and he is perfectly charmed with Payette lake.

SEARCH IN IDAHO FOR SPECIMENS.

Professor Monahan of Baker University, Kansas, and W. D. Green Start Today.

To Explore Valleys and Rivers and Travel Over Mountains to Gather Animals, Insects, Birds and Plants—May Be Out for Months.

Professor E. P. Monahan, instructor of zoology at Baker university of Baldwin, Kan., and W. D. Green, a student of the university who will graduate next year, have come to Idaho to do research work for the institution. They have secured a permit from the state game warden to collect specimens of birds and animals they may find in the state. They will also secure specimens of plants and insects. They have been in Boise nearly a week and will leave this morning for Robie creek, where they will camp for a number of days before starting out on their work.

Their plan of operations will be to begin their work in the southern part of the state and work north, taking all the summer and probably late into the fall before they will secure the specimens they want.

The university wishes to secure specimens of animals, birds, plants and insects peculiar to the region between the Rocky and Cascade ranges of mountains and Idaho was selected as being the typical location for this work.

Professor Monahan will devote nearly all of his time to gathering specimens of animals and birds while Mr. Green will give most of his time to the plants and insects. They will travel on horseback with pack horses and will leave no section of the state unexplored. They will explore the valleys and rivers, will ride the sagebrush plains and travel over the mountains. No part of the state will be too wild or remote from human habitation for them to visit it. Weeks and perhaps months will be spent in the remotest recesses of the mountains and an effort will be made to secure specimens of every living creature that inhabits them. Nothing will be overlooked. Even the mischievous house mice will have an important place, "as the mouse is just as important scientifically as the moose," said Professor Monahan.

The data that will be gathered by Professor Monahan will be of a very definite nature. At each place where a specimen is found, whether it be of an animal, bird, insect or plant altitude, latitude and general roundings will be noted. Many tographs will be taken of the p where specimens are found.

After the research work is pleted Professor Monahan will pu a bulletin showing the results of summer's work and will submit the game warden and state univer He will also leave a bulletin with of them for future use.

Baker university is the oldest lege in Kansas and many people known in Boise are graduates Dr. Dalton, pastor of the First odist church, was graduated fr university, as was also Dr. I. F. who was formerly pastor th church. T. C. Bacon of Bo is student of the university an ill be graduated next year.

Here's to your health happiness'—DeWitt's Little Ear Risers—famous little pills. Sold ll druggists.

Boise vs. Weiser, Ba Sunday, 3 p. m.

Although this newspaper clipping pasted inside the cover of one of Monahan's old field journals didn't mean much to present day biologists at Baker University (since his name appears nowhere in his journals), they immediately identified the journals for the author of this book. There is no date on either clipping but Powers' research found it in the May 25, 1907 issue of the Idaho Statesman newspaper. It provides a time-capsule glimpse of the expedition on the day they left Boise to begin their nearly month-long encampment on Ruby (Robie) Creek. The shorter note (top left) is from the Lardo (McCall) paper, ca July 5th after their arrival at the Payette Lake – Lardo area.

The two bird guides thought to be used by Monahan on the 1907 expedition: Reed's 1906 guide (smaller) and Coues' 1903 large, two volume set.

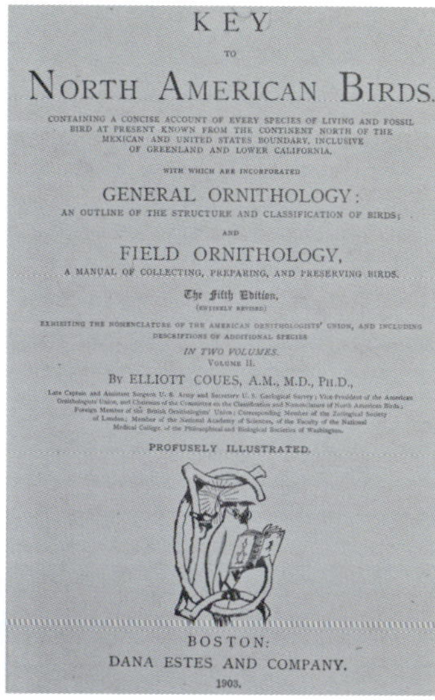

KEY

TO

NORTH AMERICAN BIRDS.

CONTAINING A CONCISE ACCOUNT OF EVERY SPECIES OF LIVING AND FOSSIL BIRD AT PRESENT KNOWN FROM THE CONTINENT NORTH OF THE MEXICAN AND UNITED STATES BOUNDARY, INCLUSIVE OF GREENLAND AND LOWER CALIFORNIA,

WITH WHICH ARE INCORPORATED

GENERAL ORNITHOLOGY:

AN OUTLINE OF THE STRUCTURE AND CLASSIFICATION OF BIRDS;

AND

FIELD ORNITHOLOGY,

A MANUAL OF COLLECTING, PREPARING, AND PRESERVING BIRDS.

The Fifth Edition,

(ENTIRELY REVISED)

EXHIBITING THE NOMENCLATURE OF THE AMERICAN ORNITHOLOGISTS' UNION, AND INCLUDING DESCRIPTIONS OF ADDITIONAL SPECIES

IN TWO VOLUMES.

VOLUME II.

BY ELLIOTT COUES, A.M., M.D., PH.D.,

Late Captain and Assistant Surgeon U. S. Army and Secretary U. S. Geological Survey; Vice-President of the American Ornithologists' Union, and Chairman of the Committee on the Classification and Nomenclature of North American Birds; Foreign Member of the British Ornithologists' Union; Corresponding Member of the Zoological Society of London; Member of the National Academy of Sciences, of the Faculty of the National Medical College, of the Philosophical and Biological Societies of Washington.

PROFUSELY ILLUSTRATED.

BOSTON:

DANA ESTES AND COMPANY.

1903.

This Bald Eagle taxidermy mount (photo by Roger Boyd) is thought to be the eagle Monahan collected in Idaho in 1907. Its strikingly similar posture to the eagle illustration in Coues' 1903 book (below), Monahan's bird reference for the expedition, lends further credence to this belief. In the eagle mount's original state, the tail would have been more horizontal (as in the Coues illustration); in time these attached tails often work loose and droop as with this old mount.

FALCONIDÆ — BUTEONINÆ: EAGLES. 697

(or more); extent 6 or 7 feet; wing 2 feet (♀) or less (♂); tail a foot, more (♀) ... Three years are required to perfect the white head and tail of the "bald" Eagle.

FIG. 474. — Bald Eagle. (From Tenney, after Wilson.)

...r, the young are "black" Eagles: very dark colored, with fleecy white bases of ... showing here and there; bill black; iris brown; feet yellow. Next year, they are

Dr. Elmer Perry Monahan while serving in the Army Medical Corps ca 1918. Stationed in Ft. Sill, Oklahoma battling the spread of the deadly pandemic Spanish Influenza, Monahan survived his own infection by the disease. World-wide, the Spanish Flu outbreak took a toll of human lives many timers greater than the casualties of World War I.

Monahan with his wife Jessie, Ca 1953.

Monahan and son at a hunting cabin in Colorado, ca 1961.

Monahan with his purebred Hampshire sheep he raised as a hobby at his rural Kansas estate, 1939.

MONDAY, JULY 29 found the men up at 7:30. This was the long-awaited day of their friend's arrival. In starting their morning fire they discovered they were down to only two remaining matches. Green literally struck out with both matches in his fire-starting efforts and so the men headed to town for matches, butter, bread, and coffee. On their return to camp they searched for huckleberries but found none. They did, however, discover lots of service berries and found them quite to their liking.

The rest of the morning passed at a leisurely pace and the men delayed lunch until after they picked up their arriving guest. Ray Green arrived as scheduled on the 1:15 train. Monahan thought his ol' college buddy looked the same as ever but "...very tender-footed."

The three returned to camp in a rather festive mood and enjoyed a celebratory feast of bread, butter, fried potatoes, prunes, crackers, honey, bacon and coffee. Monahan notes that the elder Green had brought them a 30-30 rifle from Boise, but expressed a little disappointment: "...nothing to drink as he is a teetotaler."

Eager to show their guest some good fishing, the three hitched up the old "Science Car" and headed back to their former Mud Creek camp site. They had barely traveled two miles when they realized their dog was missing. Will back-tracked to Evergreen where he soon found "Bug Head" hanging around town. After a good disciplining that would probably not qualify Green as the "Dog Whisperer", the men and their dog were soon back on the trail.

Once they hit Mud Creek, Monahan and Ray Green hopped off the wagon and began fishing while Will took the wagon on to their campsite and unhitched the horses. Mud Creek held true to its prior fishing success and the two men caught a half dozen nice trout for their supper.

The men built a big fire that evening and enjoyed catching up on a backlog of visiting and news. Monahan and Will indulged in their practice of an evening smoke. Apparently, Ray Green not only was a teetotaler but didn't smoke either; or as Monahan describes the elder Green, "...has no bad habits whatever."

The men visited late, and the night was filled with laughter and banter about roughing it while camping. The older Green snatched away his younger brother's pillow claiming it had no place in camp life – of course, in big brother humor he soon appropriates that same pillow for himself and his own comfort. There emerges a fresh sense of camaraderie among the three, meshing "brother bond" of the Greens with the reunion of former classmates, Monahan and Ray. Ray's arrival probably infused a fresh diversion into the three-month-long dynamics of a professor-student relationship between Monahan and the younger Green.

Monahan's diary notes poke fun at some of Ray Green's city slicker habits, such as insisting on combing his hair, and carrying a white handkerchief, contending in his diary: "Will have to cut that out soon." Too, there are light-hearted disagreements about cooking fish: "Ray says 'heads on', Monahan and Will say, 'take them off!' Ray says 'scales and all', Monahan and Will say 'scrape 'em off!'"

TUESDAY, JULY 30 That morning the threesome tried fishing but had such poor luck that they soon loaded their gear and began the trek back to Payette Lake. They stopped in Meadows and purchased ammunition for their 30-30 Winchester that Ray brought for them. In those days, ammo was 80 cents per box of shells. While in town, they also purchased a pack saddle and replenished their syrup supply. The pack saddle was likely in preparation for their planned upcoming hunts for big game animals. A few miles out of Meadows, they stopped for the day and located their camp beneath a pair of big pines. The weather in the Meadows area continued to be hot making the shade of the towering pines very inviting.

Shade, good grass and water were always important considerations for campsite locations. Here at the big pines, water initially posed a problem. All they could find was slough water which Monahan contended gave coffee the taste of cows; but before long Will scouted out a nice spring with good water and the camp settled into its comfortable routine. That night the men built a warming fire, visited a while and called it a day.

Their campsite in this new and unfamiliar setting hosted some unexpected visitors. During the middle of the night cattle wondered into camp and twice Monahan got up and ran them off. Yet along with this nuisance in the night, there was also music in the night – an owl serenaded the camp for awhile. There are no clues as to what kind of owl they heard calling, but our present day expeditions in the Meadows area have found great horned owls there, as well as conclusive evidence for the presence of great gray owls. Though we've not personally observed them in that precinct, long-eared, northern saw-whet, flammulated and western screech-owls could also likely be resident to the Meadows valley area.

WEDNESDAY, JULY 31 Despite their interrupted night, the men were up in time to have a breakfast of corncakes, syrup and coffee, and be started up the hill by 8:30. The day was destined to be hot.

The old wagon rumbled up the grade and four hours later they rolled into Lardo. They stopped briefly to pick up the mail then proceeded to their old lakeside camp for lunch.

During their week's absence a lot of new campers had arrived around the lake. One of the new campers, from California, had picked a good supply of huckleberries, so the Baker men decided to go out and gather some for themselves. Apparently the picking was good; later that night they enjoyed two delicious huckleberry pies for supper.

The mid-summer influx of campers around Payette Lake was not uncommon even back in those early days. Some of the well-to-do folks in the Boise Valley made a practice of escaping the valley heat by heading up into the cooler mountains. This early August flurry of campers noted by Monahan was probably a reflection of that growing tradition.

After their berry picking, the men visited their rancher friend, Gaekel, and retrieved their camp gear which he had stored for them. Monahan and Ray went into Lardo for more provisions, and sent off a letter to their Boise friend, Thomas C. Bacon, their gracious host upon their arrival in Idaho. They summoned him to come and join them the following Monday.

Not much else is recorded for that day except that their lakeside camp still held its undiminished hordes of mosquitoes. Again, they exchanged their insect pestilence. In the Meadows area, it had been flies with no mosquitoes; here it was again blood sucking mosquitoes, absent the flies.

Mint

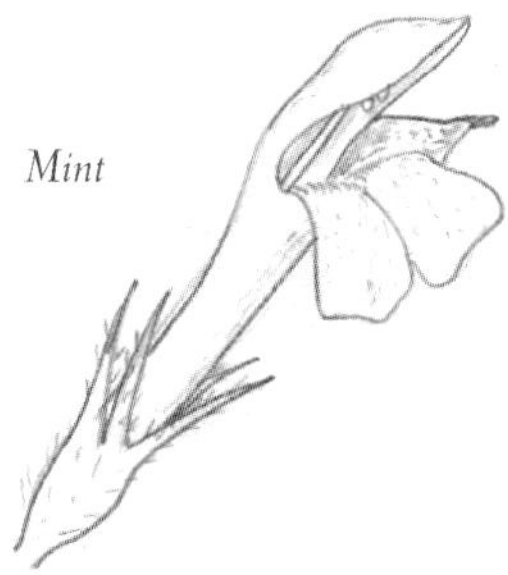

·10·

BACK AT THE LAKE

(AUGUST 1–AUGUST 6)

THURSDAY, AUGUST 1 Their first full day back at their Payette Lake camp was spent fishing. It was now nearly three months since Monahan and Will Green had left Kansas. Ray Green was eager for a boat ride around the lake and a day spent fishing. To accommodate their guest, Monahan borrowed a boat from an old Dutchman friend. The boat was not in the best of condition, as Monahan notes: "…an old leaky tub and started wondering whether we would ever get back dry and alive."

Eventually they beached their boat leaving their guns in it and did some serious fishing for about three hours. By the time they returned to the boat around 5:00 that afternoon they had caught 130 brook trout. On their boat ride back toward camp, Will rowed the boat while Monahan and Ray worked at cleaning all the fish. They also shot a "hell diver" on their return trip. Of course, this bird could have been any one of the several grebe species resident to the lake area. Fortunately, the field journal listing below resolves the grebe identity.

To no surprise, supper that evening was fried fish – two skillets full – along with home-made biscuits. Feeling well fed, the men had a good smoke and went to bed. With the mosquitoes so bad, sitting around the campfire was probably not as inviting as retreating to the somewhat less mosquito-ridden tent.

That night it grew cold. Apparently, with Ray's arrival the men had divided up the blankets and their tarp for sleeping accommodations. By dawn all three were uncomfortably cold and were forced to consolidate their bedding covers for warmth. To make matters worse, as on their last night in Meadows valley, range cows wandered into camp and had to be driven away.

~~~~~~~~~~~~~~~~~~~~~~

(Birds) *8/1 No. 85 Perisoreus canadensis (juv) (Canada* [Gray] *Jay*
Boise co. Idaho

8/1 No. 86 Perisoreus canadensis (juv) (Canada [Gray] *Jay*
Boise Co. Idaho

8/1 No. 87 Sialia arctica [currucoides] (Arctic [Mountain] *Bluebird)*
Boise Co. Idaho Common

8/1 No 88 Sita aculata [Sitta carolinensis] juv Slender-billed Nuthatch [White-breasted Nuthatch]
Boise Co. Idaho Common

8/1 No. 95 Podiceps auritus californicus [Podiceps nigricollis] (Am. Eared Grebe) [Eared Grebe]

8/1 No. 96 Accipiter velox [striatus] (Sharp-shinned Hawk)

~~~~~~~~~~~~~~~~~~~~~~

Of the above six specimens recorded as collected on that day, only the eared grebe ("hell diver") is noted in the diary account. The irregularity in numbering of the other five specimens is still not clear to me.

One of the recurring notes regarding neighbors at the lake-side camp relates to the Baker team's Dutch acquaintances. The Gaekels are prominently mentioned and seem to be very friendly and neighborly ranchers. However, from time to time in the diary there appear derogatory remarks about a "Dutch kid" that seems to be somewhat of a nuisance if not an outright pest. After their vexing night of interrupted sleep – from marauding cattle and frigid temperatures – the men are awakened early by the Dutch kid offering fish for the camp breakfast.

FRIDAY, AUGUST 2 As the day begins, one gets the distinct feeling that the Dutch kid's unexpected arrival cut short some sorely needed rest time for the sleep-deprived Baker team. Even later, after the men were up and breakfast over with, when Monahan went over to the Dutchman's place to return the boat and to hobnail and do some repair on Ray's and Will's boots, the diary records, "D----kid aggravated life out of Cot (Monahan) by his jabbering and tinkering." The lad must have been truly annoying to elicit such a response from the engaging and unflappable Monahan.

The men washed clothes that morning and laid them out to dry along the lakeshore. To help remedy the previous night's uncomfortably cold sleep, they gathered fir boughs and laid them down for insulation beneath their bedrolls. Later that afternoon, while Will went on a search for their horses, and gathered the dry laundry, Monahan and Ray hiked into town for milk and other provisions.

All this was preparatory for their trip the following day to Lake Fork, to fish that tributary of the Payette River. A neighboring camper had recently returned from that area with some impressive fish. Ray Green was obviously an avid angler and it's equally clear that one of his reasons for joining up with the Baker team that summer was for the opportunity to catch some big trout and, with luck, perhaps salmon.

That night, the fir bough mattress did the trick and the men slept comfortably and warm; in Monahan's words, "...slept like logs."

The only specimen recorded anywhere for the day was a Jumping Mouse.

(Mammals) *8/2* *N. 14* [Western] *Jumping Mouse [Zappus princeps]*
Notes as in # 13

SATURDAY, AUGUST 3 The men were up early, anticipating their day of fishing. They were on their way to Lake Fork by 8:20 – albeit on foot. In his search for their horses the day before, Will had drawn a complete blank. The horses were nowhere to be found. Not to be deterred however, the three fishermen-to-be set out on their 7-mile hike to Lake Fork. Ray Green, being the tenderfoot of the group apparently voiced some displeasure about the long walk.

The three reached their destination shortly after noon and immediately launched into what became a fishing bonanza. Will made the first catches with two large trout. He had barely landed them when his brother Ray caught a 3-pounder. Monahan added to the excitement when he lost his hook to a lunker which he contended in ecstatic hyperbole, "...would weigh at least 75 pounds!"

The fishing frenzy continued unabated until about 3:30 when the men finally grew weary of the sport. Even after gifting five trout to some local sheepherders, their final count was still 35 great trout – as many as they cared to carry all the way back to camp.

After such extraordinary fishing success, the long hike home didn't seem nearly so hard. Even Ray, who had complained some about the long hike down, was in much better spirits on the return trip.

Ever on the lookout for specimens, the men flushed a Wilson's snipe which Monahan shot as they crossed a big meadow. Of course the skin was destined for the Baker Museum, but this robin-sized, long-billed shorebird bore enough resemblance to the Woodcock of the Eastern U.S. that the Baker team planned on eating the meaty carcass at breakfast the next morning. Plovers and sandpipers were not uncommon fare for gunners back in those days.

A little further on, the men were pleasantly surprised when they happened onto their wayward horses, which had eluded them the day before. Will promptly caught them, mounted one and led them back to camp. I don't know if Ray Green, earlier alluded to by Monahan as a "tenderfoot" was uncomfortable with horses, or he simply didn't want to abandon his ol' Baker classmate to walk alone back to camp, but for whatever reasons, the weary Monahan and Ray Green continued on foot after Will departed on horseback for camp. This, even though Ray was still suffering from blistered feet. Apparently, Monahan's work on the Green brother's boots the day before hadn't cured all ills. Maybe the annoying Dutch kid had driven the would-be cobbler to absolute distraction.

By now mosquitoes were eating the two men alive. But the discovery of a thick huckleberry patch soon dulled their misery as they fairly stuffed themselves. They must have tarried a good deal at the huckleberry patch, for by the time the two older men eventually arrived back at camp, Will was a little miffed at their tardiness. He had been quite industrious since his return with the horses. While awaiting his companion's return, he had cleaned the fish and fried-up some along with corncakes for supper; and, carried in water from the spring.

Exhausted from their 14-mile round trip hike and incredible day of fishing, the men were famished and ate heartily. After supper the Green brothers took the "Ol' Science Car" into town to check on word from their Boise friend, Thomas C. Bacon. Monahan remained in camp with Bug (the dog) and skinned out the snipe. The combination of retrieved horses and weary men no doubt accounted for the hitching up of the wagon for the trip to town.

~~~~~~~~~~~~~~~~~~~~~

(Birds) *8/3 No. 83 Gallinago wilsoni [delicata] (Wilson's Snipe)*

Taken on Payette Lake
~~~~~~~~~~~~~~~~~~~~~

8/3 No. 84 Podeiceps auritus californicus [Podiceps nigricollis] [Eared Grebe]

Taken on Payette Lake Boise Co. Idaho

8/3 No. 97 Asyndesmus torquatas [Melanerpes lewis] (Lewis Woodpecker)

Lake Payette, Boise Co. Ida

(Mammals) *8/3 No. 15 [North American] Porcupine [Erethizon dorsatum]*

Taken in high mountains near Payette Lake Boise Co. Ida.

~~~~~~~~~~~~~~~~~~~~~~

Of the above four specimens recorded in the field journals for this date, only the Wilson's snipe is noted in the diary account. Contrary to its notation in the field journal, the diary describes the snipe being flushed out while crossing a meadow some distance from camp, rather than taken on Payette Lake as is stated in the field journal. I suspect this is simply a difference in tenor of narrative details employed in the twin records: diary versus field journal description.

Interestingly, these are the last specimens recorded either in the diary or in field journals for over a month. It wasn't until September 5 that Monahan finally again records any collected specimens and those are several small mammals. No bird specimens are again recorded until September 9, even though Ray and Bacon leave on August 21. I suspect two factors play strongly into this hiatus of specimens. After the departure of their visitors, Monahan and Green concentrate first and foremost on big game hunting. Plinking around the countryside knocking off small game specimens during those hunts would prove counter productive from the standpoint of spooking their principal quarry – big game. Then, later on, Monahan's predicament with a vexing skin rash kept the team out of the collecting mode for an additional 10 days.
~~~~~~~~~~~~~~~~~~~~~~

SUNDAY, AUGUST 4 Like so many of their previous Sabbaths, this was a quiet day in camp. They baked three large huckleberry pies and ate one for lunch. The weather had turned cool, so much so that Ray Green donned a sweater.

During the afternoon Will and Ray wrote letters to family and friends back home while Monahan worked on bird skins. Earlier that morning Monahan had shot a nuthatch that made the fatal error of flitting about their camp. This particular nuthatch is not identified by species, so we don't know whether it was a red-breasted, white-breasted or pygmy nuthatch. A week earlier while in the Meadows area, they specifically mentioned seeing a pygmy nuthatch, but make no mention of collecting it. A white-breasted nuthatch is recorded above (August 1) in the field journal. With the journal specimen numbers and dates appearing in somewhat mixed array, it is even possible that this diary record alludes to the August 1 bird which is not mentioned in the diary account on that date.

That evening Monahan makes another tantalizing diary entry when he reports developing pictures. Oh, wouldn't it be grand to have some of those photos taken on this 1907 expedition! Regrettably, they still elude us, if they exist at all.

That night Monahan recorded temperatures down to 36° F, and noted, "We slept under four pairs of blankets & were none too warm."

MONDAY, AUGUST 5 This was the day their Boise friend, Thomas Clyde Bacon, was to join them. With that anticipation, the Baker bunch arose early, had breakfast and went in to town to check the mail for word from Bacon. There was nothing new from their expected guest but while in town they bought a pair of "alfarkosn" saddle bags, some angling flies for their fishing gear, and collected some old newspapers for the plant press.

The last item – old newspapers for the plant press – begs a couple of questions. First, why is there absolutely no mention of the kinds of wild flowers collected and put into the plant press? With the aforementioned *Gray's Manual of Botany* being such a valued treasure of Monahan's, and most likely carried along on this expedition, one would assume that there would have been recognizable plant varieties encountered and hailed like old friends in the diary record. But there are none! I assume the presence of a herbarium (cabinets of plant specimens) as part of the old Baker Museum, but I have absolutely no recorded proof of such.

The other question relates to the contents of the old news papers that were being put into the plant press. What might have been the breaking news of the day – Idaho in general and Long Valley in particular? In late July of 1907 Idaho was creating quite a stir in the national spotlight. The "Big Bill" Hayward trial, remembered as the country's first "trial of the century", was just then winding down in Boise, Idaho. This dynamic trial of the alleged assassin of former Idaho Governor Frank Steunenberg pitted the renowned defense attorney, Clarence Darrow, against William Borah, Idaho's silver-tongued debater and newly elected senator. Certainly, any news print would have been awash with coverage of this nationally watched trial. If that drama didn't play out in the Long Valley newspaper, Ray Green would have encountered the buzz while passing through Boise. Surely, Thomas C. Bacon would carry some of that political news with him when he arrived later that day. If so, none of this trickled into Monahan's diary account.

Locally, Lardos' *Long Valley Advocate* would have contained the news of the startup work on a new flour mill in Roseberry. Some farmers mid-way down the valley were reported having trouble with sheepherders. The Deputy Game Warden that had made no issue with Monahan for being unable to immediately produce his collecting permit made the news on another matter involving a misunderstanding about some fishing violations and fines imposed by a local justice. The local paper mostly printed local chatter about various valley residents going here and there: a baseball game at Roseberry between the home team and Thunder City; and a parent's meeting at the local School District 38 schoolhouse to devise means for better school attendance. If the Baker team ever read the paper's notice about the Hotel McCall's Sunday special chicken dinners and ice cream suppers, they apparently never indulged in such treats.

That afternoon the men again took advantage of the local bounty of ripening huckleberries. The rewarding I ½ gallons of berries picked seemed to dull the misery of the accompanying hordes of mosquitoes. Sometime during their outings Monahan and Ray each had shots at a coyote but both narrowly missed.

Later that afternoon Monahan developed pictures he had taken that morning, but discovered they were no good. Here again, this stated difficulty in getting photographic results is perhaps testimony to the reason for no surviving photos from the expedition.

Thomas Bacon finally arrived that evening – presumably by stagecoach – and the three campers went in to meet him. Unlike the teetotaler Ray Green, Bacon brought with him some wine, along with fresh apples. In a festive spirit the four Baker alumni returned to camp, built a big campfire and sat around swapping stories, singing and...well, the revelry is best summed up by Monahan's succinct diary notation: "... good time."

TUESDAY, AUGUST 6 Tacit affirmation of the previous night's late partying and 'good time' is this day's diary entry – one of the three shortest notations of the summer: "Took George a boat ride. Spent rest of day visiting and talking over old times."

In keeping with their practice of assigning aliases to each of the expedition members, Thomas Bacon assumed the alias, "George Cole", a former Kansas State Auditor. Ray Green's alias, acquired soon after he joined the team was Mort Albough, a former Kansas newspaper editor, Republican State Chairman and State Bank Commissioner. Socolofsky (decipherer and publisher of Monahan's diary) notes that of the four aliases used by the Baker team, only one figure, Arthur Capper, actually succeeded in their political careers. Perhaps this was their motivation for bantering the political names – tongue in cheek humor – poking light-hearted fun at their somewhat rapscallion state politicians.

Though the day seemed relaxed and uneventful – the only outing of the day involved a boat tour of part of the lake with Bacon – very likely the men also tended to some details of breaking camp the next day. They were gearing up for a four-day trip east to the rapids of the Lake Fork River. And from there they planned to trek further east-northeast to Lick Creek for an additional week. June and July had come and gone and Monahan was growing increasingly impatient to collect some big game animals. After some quality fishing on Lake Fork, the team planned to head up into the high country, hoping for improved chances of finding deer, elk and bear. They would end up traversing alpine zones that could also harbor mountain goats.

Silene

·11·

TO LAKE FORK AND LICK CREEK AFTER BIG GAME

(AUGUST 7–AUGUST 21)

WEDNESDAY, AUGUST 7 The Baker team arose early and began breaking camp, in preparation for departure on their Lake Fork and Lick Creek trek. After breakfast they hauled some of their unneeded camp gear over to the Gaekel farm and left it there for temporary storage. Because they were bound for some rugged trails up into high timber in search of big game, their wagon would be of no use to them. Instead they'd travel on foot, using the two horses as pack animals for carrying their camping essentials.

In his diary that day, Monahan gives a rare and briefly detailed description of his three fellow trekkers as they departed the lake that morning. Thomas C. Bacon wore a green shirt, Khaki pants and a Stetson hat. Ray Green's apparel was a dirt-straw colored shirt, a faded pair of overalls with sack strings for suspenders, and a black hat which Monahan contends is entirely out of place. Will Green was still wearing his coveted white felt hat, a striped shirt of "indescribable color", and Khaki pants. All of the men wore high-laced boots. In characteristic fashion, Monahan modestly makes no mention of himself or his own apparel.

Their food provisions for the two-week excursion are listed as: bacon, tomatoes, potatoes, flour, beans, butter, coffee, tea, prunes, and onions. Of course, they surely planned to supplement these items with fish and wild game acquired along the way.

Traveling on foot, single file, one man leading a horse while the other carried weapons, the men marched steadily along and reached the Lake Fork rapids by mid-afternoon. There they set up camp and soon went about fishing, catching enough fish to supplement a Dutch oven mulligan cooked up by the Green brothers. The men feasted ravenously that evening.

Their fit and fettle mood was somewhat deflated when about bedtime they discovered that their horses had again wandered away from camp and apparently headed back to the lake. That would mean a long hike for someone tomorrow to go back and retrieve them, disrupting their plans.

THURSDAY, AUGUST 8 That morning the men arose feeling hale and hardy after their previous day's seven mile hike from Payette Lake. Monahan and his loyal dog Bug(head) soon took to the trail, heading back to Payette Lake to find and retrieve their wayward horses. The other three men remained at the Lake Fork camp passing the time fishing and hunting.

Eventually catching up with the truant horses, Monahan borrowed a saddle from a local rancher by the name of York, and opted to ride on to Lardo for fresh supplies. In addition to some tomatoes and potatoes, the expedition leader purchased a new pair of Khakis for Ray Green, whose faded overalls were wearing thin at the seat - or as Monahan's diary tacitly states: "...getting rare in the rear."

Also growing concerned about the wellbeing of his dog, which had eaten no breakfast and was near famished, Monahan rode on over to the Gaekels for some milk for his dog. As luck would have it, he ended up with a meal of his own, accepting the Dutchman's invitation to tarry and have lunch with them. One can't help but wonder if this may have been part of Monahan's underlying intentions.

Meanwhile back at the Lake Fork camp, the three younger expeditioners were not faring so well. Despite their best efforts, fishing was dismal. Nor were they able to rustle out any new specimens for the museum collection. By the time Monahan had returned he found the men tired and rather dejected. In that day's diary notes, Monahan invokes a little of his religious roots when high-lighting the trio's fruitless take of fish and specimens, in the words of the old church hymnal refrain: *"...no not one."*

But alas, the sure cure for a futile day in the wilds is a hearty meal. That evening a Spanish Mulligan combining ten ingredients – potatoes, tomatoes carrots, turnips, beets, beans, rice, butter, onions and bacon – simmered in their Dutch oven. The meal scored high as a gastronomic delight, and the camp's gloomy mood wafted away with the meal's waning aroma.

Stuffed and now more contented with the day, the men sat around the fire, smoked, and talked of astrology, astronomy and philosophy. One can well imagine the rigorous discussions that ensued, as summed up by Monahan's tongue-in-cheek conclusion to the day's diary notes: "...settled some big questions."

FRIDAY, AUGUST 9 Early that morning while sleeping under the stars, the team was rudely awakened by a passing shower. They scrambled to erect their tent, and as so often is the curse of mountain squalls, by the time they got the tent up, the rain had stopped. Bacon seemed to have suffered the worst from the unexpected drenching; and the diary also got a brief baptism, but fortunately suffered no real damage.

Since the main purpose of this Lake Fork junket, besides the good fishing, was to strike up into the higher country in search of big game and other new specimens, the men loaded their horses with their camping gear and began the climb to higher elevation. The trail they followed most of the day was a fairly typical mountain trail, though sometimes a little challenging for their "fossilistic" horses. At one point while crossing a stream, old Bally slipped and went down, and had to be unloaded before they could get him back up on his feet. Luckily, he survived without injury and about 4:00 o'clock in the afternoon the expedition found a small island on which to camp. Close by was good grass for the horses.

The diary notes that earlier in the day, along the trail, they had encountered a family of grouse – a hen with six young – all of which were collected. There is no mention of which grouse species these were; neither reference to fool hen or ruffed grouse. Nor does the field journal record the taking of these grouse. Perhaps they were destined solely as camp meat.

There is no other mention of specimens collected that day, but while Monahan and Bacon were setting up camp, the Green brothers went off fishing. Almost immediately Ray came dashing back to camp all excited, announcing that they had three large bull trout cornered and needed help. Monahan quickly fashioned a 'snag hook' out of some large fish hooks and the men spent the next half hour trying to gaff the large trout. The fish however, were too wily and the men were unsuccessful. The Green brothers turned their attention back to their regular stream fishing with notable success; catching 30 brook trout.

While the men were engaged in their snagging assault on the cornered bull trout, a passing rain shower once again caught them unaware, soaking them thoroughly as well as their camp and guns. Traveling light and camping rough, conditions were crude and makeshift and certainly more vulnerable then usual to such whims of weather.

SATURDAY, AUGUST 10 The men slept unusually late, until 8:30. The colder air of higher elevation likely accounted for their sleepless night and complaints of cold feet in the morning.

Breakfast consisted of coffee, graham gems and lots of fish. In a light-hearted wager, Thomas Bacon ate 12 fish and suffered no ill effects. Perhaps his hearty trout breakfast even helped him muster the finesse to throw his touted 'diamond hitch' on the horse packs that morning. With horses loaded the expedition headed up the trail around 10:00 A.M.

The men climbed ever higher into the mountains that day. At first the trail was good and they passed through three bands of sheep, approximately 2500 each. As they approached the snowline, the going got rougher and the trail perilously narrow. At times Monahan worried that they might lose a horse over the precipice to the rushing stream several thousand feet below. Monahan estimated their elevation at 10,000 feet, although maps today show higher peaks in the area to be closer to 9,000 feet. They passed snowfields and stopped for lunch in a small alpine meadow where they could rest, unpack the horses and let them graze.

After this midday respite they again loaded the horses and continued on toward Lick Creek. Along the way they saw encouraging signs of deer and elk tracks. In contrast, the waters of this high country seemed increasingly devoid of fish, a condition Monahan attributed to the high elevation. Though there is no mention of specific fishing stops, it is quite likely that the team passed by small lakes and headwater tributaries feeding down into the Lake Fork River. It would have been nearly impossible for the men to bypass these sources without casting a fly or two, testing the waters.

Late that afternoon, as they approached a patch of timber there was a brief flurry of excitement when Ray Green spotted a bear. But the bruin disappeared before anyone could get a shot at it. The diary notation doesn't define the bear as grizzly or black bear. Although the latter would be the more likely, in 1907 there could still have been grizzlies roaming that part of Idaho, especially in such remote high country. A mere two decades earlier there had been a documented presence of a grizzly down in the valley near Smiths Ferry (Woods 2002).

With hunting spirits buoyed by both signs and sightings of big game, the men trekked on until finally locating a camp spot along the banks of Lick Creek. Here they stationed their base camp for several days, focusing their efforts on combing the surrounding area for big game.

SUNDAY, AUGUST 11 This was a day of sorely needed rest. The men and their pack horses were weary from yesterday's climb through the rugged high country. No better day than this back country Sabbath for some welcomed rest.

Monahan and the elder Green relaxed by spending part of the day fishing. Meanwhile, Will Green and Bacon did some scouting nearby for a more suitable base camp site. Fishing was good and the team enjoyed another supper of 30 fish.

Following supper the men sat around the fire and laid plans for the next day. A new campsite had been found; but rather than tie up the whole crew moving camp, they decided that Monahan and Will Green would set out and hunt for deer, while the other two men relocated camp.

MONDAY, AUGUST 12 The renewed men were up early, anticipating an active day in the high country. Soon after breakfast all set about their respective tasks, Monahan and Will scouting for deer, Ray Green and Bacon relocating their camp to the new site and then rewarding themselves with a catch of 50 fish.

Monahan and Will were not so successful. For all their efforts they saw only tracks of the deer they sought. On their way back to camp, however, they did happen onto a fool hen. Reluctant to fire their guns at a mere grouse for fear of scaring away big game, Monahan easily dispatched the grouse with a rock. The bird's tameness, sufficient to be killed with a stone – hence, "fool hen" – is more typical of the spruce grouse than either of its mountain cohorts, the dusky or ruffed grouse. Like three days earlier, when a family of grouse was supposedly taken in a similar fashion, Monahan's field journals record nothing of this grouse collection. Though the grouse skin may never have made it to the Baker Museum, no doubt the grouse meat provided a welcome break from fish for their evening's supper which Monahan notes "...all ate like fiends."

After supper, the men sat around the campfire and played the card game, pitch. This is the first reference to the men's playing cards throughout the entire summer's diary account. Perhaps the two additional guests made for more spirited gaming; or maybe the high country reprieve from mosquitoes granted them more freedom for some fireside revelry?

TUESDAY, AUGUST 13 Though all four men rose early, they spent the day in somewhat leisurely pursuits, mostly hanging around camp. After breakfast the Green brothers volunteered for huckleberry duty. Monahan and Bacon remained behind to tend to camp chores, and catch up on other projects. Bacon, who in the fall of 1907 would serve as president of the Biblical Society on the Baker University Campus, worked on his inaugural address.

After doing the washing, Monahan attended to other mundane expedition tasks. First he finished preserving the fool hen skin he had collected the day before. Then he placed some area wildflowers into the plant press. Again, even though the diary records the collection of plants, there are no specific references to any by name, common or scientific. Nor do the field journals record any botany specimens. Perhaps for efficiency sake Monahan jotted the wildflower data directly onto the plant press papers that sandwiched each specimen. That way, later when mounting the specimens on herbarium sheets, the collection data would still accompany each sample, eliminating the need for an additional (third) field journal.

In late summer the mountains landscapes around Payette Lake abound in wildflowers. Based upon some of my own observations, if we could somehow peer inside Monahan's plant press, the collected specimens might include many of the following: asters, bear berry, bear grass, bedstraw, butter-and-egg plant, checkermallow, western chokecherry, ciliate bluebell, elephant-head, goatweed (St. Johnswort), goldenrod, green gention, Jacob's-ladder, Indian paintbrush, Lewis's monkey-flower, lupine, onion, mariposa lily, western monkshood, mountain golden pea, orange mountain dandelion, Oregon grape (creeping), parrot's beak lousewort, *Penstemon* (shrubby, hot-rock), pearly-everlasting, pink spirea, Piper's anemone, *Potentilla* (slender), Rocky mountain (yellow) pond-lily, scarlet gilia, *Senecio* (groundsel), shooting star, stonecrop, streambank globe mallow (wild hollyhock), western coneflower, western horsemint, wild buckwheat, wild rose, and wild strawberry.

Around 4:00 that afternoon Ray and Will retuned to camp with two quarts of huckleberries. It was now nearly 10 days since the expedition had enjoyed their last huckleberries at the lower elevations around their lakeside camp. The local campers, farmers and townspeople had likely depleted most of the waning valley crop but now, up in the remote stretches of higher mountains the belated berry crop was fresh and uncontested except by wild animals.

While the berry pickers rested from their roaming excursion, Monahan and Bacon began supper. That evening they dined on another 'Mulligan' served up with sour dough biscuits, syrup, huckleberry sauce and coffee. After supper Monahan and Bacon tended to the task of washing the dishes, while the Green brothers took off for some evening fishing.

Just before dark Will and Ray Green returned with a good catch of 23 trout for – you guessed it, tomorrow's breakfast (by now you've figured out that there is no shortage of omega-3 fats in these men's diets)! Upon their return they all sat around the fireside and, like the night before, played cards until bedtime.

WEDNESDAY, AUGUST 14 Monahan and Will Green woke up at 5:20 A.M., and slipped quietly out of camp looking for deer. Later that morning Ray Green arose not feeling well. After eating a little breakfast he was still feeling so poorly he returned to his bed. I would wager maybe he was "sick" of too much fish.

Meanwhile, the hunting party was not having much luck in the woods. Though they found very fresh sign, again they were unsuccessful in seeing any deer. Finally, around 4:00 that afternoon they returned to camp empty-handed, discouraged, and hungry – having not eaten since breakfast. By then, Ray Green was feeling somewhat recovered and within the hour they all enjoyed a hearty meal.

Still determined to bag big game, after supper Ray Green and Bacon took the rifles and went out for an evening hunt for deer.

They were no doubt hoping to succeed where their companions had miserably failed. Alas, they too returned to camp empty-handed.

At this juncture of our story, it is tempting to speculate about the men's poor hunting success. Were they simply not familiar with these animal's habits? Being from Kansas, were they more accustomed to hunting the Midwest's native white-tailed deer than the mule deer that frequented the Idaho mountains in which they now hunted? Thomas Bacon, however, was a Boise, Idaho native and very likely a veteran mule deer hunter. History does tell us that big game had been heavily hunted as meat-marketing support for many of the early mining operations scattered throughout the Idaho Mountains. Could that have played a factor in the Baker team's poor hunting success? This 1907 account strongly suggests the scarcity of big game.

THURSDAY, AUGUST 15 Bacon woke Monahan early and the two slipped off for an early morning deer hunt. Unsuccessful, they returned to camp at 8:30, sorely in need of some breakfast. Though the dog eagerly greeted their arrival, the two other men were still in bed.

Smarting a little from their lack of success in bagging any deer, the men set out after breakfast with pails and rifles to hike down the canyon looking for huckleberry patches; that may also be an attraction to bears. Armed with the 30-30 and 40-72 Winchesters, the 38 Iver Johnson and the .22 Marlin, I expect each man carried a rifle.

The men encountered no bears but did find enough huckleberries for their supper. When they returned to camp around 5:00 P.M. they were surprised to find two Irish sheepherders there. During this genial visit the herders offered the Baker Team some mutton. This offer no doubt promised a welcome change to the Baker team's diet of fish. It isn't clear whether or not the sheepherders stayed for supper, but the diary notes for the day concluded with "Supper over, everyone sat around the fire until bedtime."

FRIDAY, AUGUST 16 That morning Monahan slipped out of his warm bedroll at the predawn hour of 4:30 and again went in search of deer. And, as usual, he returned to camp empty-handed.

That day's diary entry reflects not only Monahan's growing frustration with their abysmal deer hunting success, but he also laments the fact that their two visiting friends are feeling the need to return to their lives outside the Idaho mountains. It has now been three months since Will Green and Monahan left Kansas. Their Idaho odyssey is more than half over.

SATURDAY, AUGUST 17 The men broke camp that morning and started back toward Lardo. I expect it may have been a melancholy journey for Monahan. The sole purpose of this 10-day junket – to collect some big game specimens for the expedition – had been a total bust. On top of that, their two visiting comrades would soon depart leaving the original twosome.

Not much else is recorded about their travel that day except that they found a grassy spot on Lick Creek to camp for the night, a spot where they caught fish and ate well for supper.

SUNDAY, AUGUST 18 Although it was a Sunday, they broke from their usual practice of resting on the Sabbath. Instead, they got up at 8:30 and completed their trek back to their Payette Lake campsite, arriving at 2:50 P.M. Nothing else is reported in the diary that day – perhaps a reflection of subdued spirits tempered by the upcoming departure of friends.

MONDAY, AUGUST 19 Still worn out from their high country odyssey, the men were content to lounge around camp resting. However, news of eight great salmon that Mr. Gaekel had recently caught quickly revived them. According to current reports in the *Long Valley Advocate* (September 19, 1907) the salmon run in the river was the best in several years and many were being caught. Evidently, the Dutch farmer's catch was part of that bountiful harvest and he shared generously with the Baker team. Seeing the huge fish, Ray Green was at once eager to go salmon fishing. That may have put the two visitor's departure plans on hold, as they immediately laid plans for a salmon fishing trip the next day.

TUESDAY, AUGUST 20 The men awoke to frost on the grass that next morning. Will Green didn't want to leave the warmth of his bedroll but the others dragged him out, anyway. Shivering in the cold, they dressed, stoked the fire, and hunkered near it for warmth and breakfast. Anticipating a day of hauling in great salmon the men feasted on fried salmon (compliments of neighbor Gaekel), oatmeal, bacon, prunes and coffee.

Most of that day was spent in quest of salmon. But, unlike their usual great success with trout fishing, they caught no salmon. After supper Monahan and Ray Green visited the Gaekel's farm to acquire some fresh milk.

WEDNESDAY, AUGUST 21 The men were up at 4:30, well before dawn. This was the morning that Ray Green and Thomas Bacon would bid farewell and board the Lardo stage at 6:00 A.M. From there they would journey to Weiser, and then on to Spokane. From Spokane, Ray Green would return to Kansas, while Bacon would travel on to Portland, Oregon to visit his mother.

It was another profoundly cold morning – "...cold as Greenland and 34 above." according to Monahan. After a hurried breakfast they hitched up the horses to the old 'Science Car' and hauled their friends and their belongings to town to meet the stage. They arrived well ahead of the stage but too soon had to say their farewells as the stage rumbled off. There is a palpable sense of sadness at Monahan's notation of their departure: "The time soon came and off they went leaving us alone and somewhat lonely too."

The Baker team found their separate diversions that day, probably helping to take their minds off their departed friends. Stopping by the Gaekel's on the way back to camp, Mr. Gaekel invited Monahan to join him and his son, Walter, salmon fishing. Will Green stayed behind at the farm to pitch hay.

On a historical note, the frosty mornings of the past two days noted in Monahan's diary are also chronicled in the early September issues of the 1907 *Long Valley Advocate*. The September 5 issue states that "the very unusual frost of several days ago" had "entirely killed" wheat in parts of the valley, and although the oats suffered some damage, there would be a fair crop. In lower parts of the valley, near Center, the August killer frost did no damage to the Barley crops; indeed some fields averaged sixty bushels per acre.

The August 8, 1907 issue of the *Long Valley Advocate* also notes a shortage of haying hands in parts of the valley. At that time the going wages for a worker was $2.00 per day. Monahan's diary does not specify any of the working terms for Will Green's help with the haying at the Gaekel farm that day. Whether it was volunteer service to spring loose the elder Gaekel and son to take Monahan fishing, or if it was paid labor is not clear. Given the neighborliness and kindnesses that the Gaekels customarily bestowed upon the Baker Team, it is very possible that Green labored on their farm for the mere wage of a home-cooked lunch and supper for the day.

Monahan and his farmer friends returned later that day with a 15-lb. salmon. It was divided evenly, with Monahan also keeping the skin for taxidermy purposes. That evening Monahan and Green remained as supper guests at the Gaekels.

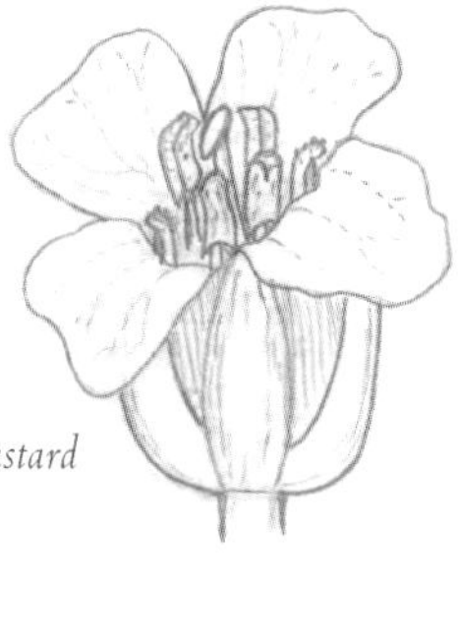

Mustard

· 12 ·

AN INTERRUPTED QUEST FOR MOUNTAIN GOATS

(AUGUST 22–SEPTEMBER 3)

THURSDAY, AUGUST 22 The original twosome were up early preparing for another big game hunting trek high into the Lick Creek country. They planned to be gone for a couple weeks, so before heading off to the hills again, Monahan wrote several letters while Will Green dismantled camp. Though it isn't confirmed in the diary, the men probably dropped off some of their permanent camp gear for safe-keeping again at the Gaekel farm on their way to mail their letters in Lardo.

Eager to be on their way, they didn't even take time for lunch. By 1:00 o'clock they had their horses heavily loaded and were headed up the trail to Lake Fork. Their eventual destination for this trip was the 'Goat Peak" area among the high granite peaks above Lick Creek. They were especially after mountain goats, but, of course, would readily take any deer, elk or bear that crossed their paths.

The going was hot and the trail was "...almost as dusty as Kansas roads..." the diary notes. Monahan walked and led Bally while Green brought up the rear with old Star Crust. They arrived at Lake Fork around 7:30, tired and very hungry, since they hadn't eaten since 7:00 that morning. The horses were no better off, and were fed oats and then turned loose to feast on the tall sweet grass that carpeted their camp spot.

The men quickly turned their attention to supper once the horses were tended. Monahan threw together his "famous" sour dough biscuits while Green produced the coffee, fish and prunes. They ate "…like fiends…", and sat around smoking and planning the next day's trek onto the high ridges for goats. Before bedtime Monahan set a trap at a nearby badger digging and gratefully sank into his bedroll. Monahan and Green were both bone weary from the day's long hike and knew that the morrow would entail climbing and scrambling among the high granite peaks in search of goats. Both men and their dog were dead asleep by 9:30.

FRIDAY, AUGUST 23 The day began slowly for the Baker team. Despite their early bedtime the night before, the weary men slept late. After checking the undisturbed badger trap, the men had breakfast and began a climb upward toward the lofty peaks above camp. Eventually, they gained enough altitude to view the nearby promontory that purportedly held mountain goat bedding areas. From their vantage point the men then spotted a ravine that would take them directly up to the bedding area. Their plan was to return that evening and ascend the peak and sleep in the vicinity of the goat beds to surprise the arriving goats in the morning.

Returning to camp, the men busied themselves with respective tasks. Monahan took the .22 and set out to shoot a 'mess' of grouse for supper. Meanwhile, Green fried out some suet, taking the rendered grease and giving his boots a good protective coating. That done, he sauntered off to do some fishing.

As might be expected, it was far easier to catch a mess of fish than find grouse in the woods, and thusly, Monahan returned grouse-less while Green brought back enough trout for supper and breakfast.

Immediately after supper the men departed for the mountain top, leaving Bug (the dog) behind to "guard" camp. It was about a mile climb steeply upward all the way. Near the top it became a scramble, hand over hand ascending granite ledges until they finally reached a suitable hideaway; it overlooked a lush, grassy clearing bisected by a clear, beautiful brook. They found this spot none too soon, for it was rapidly growing dark even at their top-of-the-world, 9000-foot elevation. There they would sleep and in the morning lay in wait for goats.

As they settled in for the night on a ledge beneath a scrub pine, each with their single blanket for a bed, a trio of discomforts began to disturb their sleep. Almost immediately, large black ants contested their presence. Soon the men were forced to relocate a short distance away. The absence of any specimens render their identity iffy but very likely those irritating insects were carpenter ants (*Camponotus modoc*). Hardly had they settled into their new spot when a small, furry, rat-like animal began annoying Green by nibbling at his toes. By then a full moon illuminated these shenanigans and the men could plainly see the animal. Tired and growing furious at the disturbance, Will kicked at the varmint several times and finally scared it off. To Monahan, the consummate naturalist, this critter represented a specimen he didn't recognize and secretly hoped it would return so that he might collect it. The pest resembled rats he had seen back in Kansas, except this version was larger and had a bushy tail. From his description it is obvious that their pesky night visitor was a bushy-tailed woodrat (*Neotoma cinerea*), known more prosaically as "packrat". Monahan likely knew the similar but smaller and slender-tailed Kansas native, the eastern woodrat (*Neotoma floridana*), but this larger, bushy-tailed northwestern cousin was a stranger to him.

It was well past mid-night by the time the woodrat episode finally subsided and let the men sleep. But alas, their third vexation began. Their steep, arduous climb to the mountain top had left the men dehydrated. That plus the surprisingly warm nocturnal temperatures on the peak sent them to the little stream several times during the night to slake their thirst.

SATURDAY, AUGUST 24 Green awakened Monahan around 4:00 a.m. and suggested they begin their vigil for mountain goats. It was still dark but the men quietly arose and positioned themselves and sat gazing ahead while their eyes adjusted to the dawn's faint glow. At the point where they could just barely discern objects around them, the men were suddenly startled by the sight and sound of a beautiful big buck bounding away. Apparently the deer caught their scent and bolted.

The surreal experience of the vanishing buck was both disappointing and exhilarating. Shadowy image that it was, the men had seen their first deer, and a trophy buck at that. It jolted them awake and bode well for their morning vigil at this mountain goat peak. However, that was to be the extent of their big game sightings that morning. No goats or any other big game animals made an appearance at their mountain top lookout. On the way back down the mountain the men saw bear tracks, which were becoming a fairly common sighting in the area. Though they returned empty-handed, the encounter with the trophy buck had nonetheless buoyed their hopes. They decided to rest the remainder of the day and return to their goat peak area with two- to three-day's provisions, enough to extend their time and allow them a greater promise of success in their hunting. They were feeling much encouraged about their chances of taking big game specimens. That day's diary account is infused with the notion that success was just one hunting trek away.

Returning to camp they found Bug doing fine but elated to see them. They had no inkling of an imminent run-in between the dog and a set of jaws. It happened soon after breakfast while Bug joined Monahan in fetching the horses. Somehow, the dog stepped into one of Monahan's # 4 1/2 jaw traps, probably intended for Lynx, cougar, or even bear. That size trap is powerful enough that it took both Green and Monahan to release the dog's foot. Fortunately, Bug's foot suffered no broken bones, but doubtless he was "gimpy" for a few days.

Rain that afternoon prevented the men from fishing or specimen collecting. The countryside and its trails had become very dry, so the rain was welcomed. With the blessing of moisture to the mountains, the men were content to spend the rest of the day sitting in their tent. They assembled an early supper and went to bed. The rain continued into the night and through the next two days, keeping them mostly confined to their tent. We Idahoans who have lived long in our state know of this late August-early September storm pattern; it commonly socks-in the mountains, wetting the lowlands and laying down a blanket of snow in the high country. We know that Monahan was aware of this weather, as he describes it in his diary: "This may be the fall rains here after which we can look for cold nights, frequent frosts and snow occasionally."

SUNDAY, AUGUST 25 During the first rainy day, a big fire built in front of the open tent door warmed them as they and the dog lay inside waiting for the rain to abate. Meals had to be hastily cooked outside in the rain, since they had no stove for their tent. The rain continued all day, through that night and throughout the next day.

MONDAY, AUGUST 26 The rain persisted. Nonetheless, the men, growing antsy, braved the wet weather for a quick fishing trip and returned with enough fish for supper and breakfast. Still, it remained cold and wet and Monahan and Green developed cabin fever. This prolonged storm thwarted their hunting plans just when they felt on the verge of success.

TUESDAY, AUGUST 27 Finally, the next morning broke clear and dry. The men scuttled out of bed and into the sunshine. They fixed breakfast and gathered provisions for their three-day hunting trip on the mountain top. Sunny skies held promise for now as the men loaded up and began their ascent. On this trip, Bug, the camp dog joined them.

They reached the top about 7:00 P.M. and located their camp in a ravine beneath some large pines. A picturesque little stream ran nearby, cold as ice and with a white-sand bottom. Monahan was quite impressed by the beauty of the camp's alpine setting. From his diary description we get the sense that this rarified outpost was to be their last stand – their rendezvous point with hunting success.

Feeling that they were once again residing in the heart of prime big game habitat, stealth became a primary concern. Thus, the men ate in silence and made just enough of a fire to boil coffee and fry bacon. They planned to use this spot as their base camp, scouting around to find where the deer ranged. Then, as Monahan confidently states in his diary, "...we can begin to hunt right."

WEDNESDAY, AUGUST 28 They were up in the faint light of daybreak. The night had been colder than they expected and neither slept well. Only the dog seemed warm enough, sleeping at their feet with two coats over him.

Stationing themselves near their camp they sat quietly and watched for deer until sunrise. When nothing showed, they ate breakfast and set out to scout and hunt through the area. Part of their reason for lingering three days in this high mountain retreat was to explore and ferret out hotspots that deer and other big game frequented. No doubt their numerous and fruitless hunting forays in the lowlands throughout the summer had convinced them that the more productive hunting lay in the high county. Their recent mountain top encounter with the trophy buck probably reinforced that notion.

The two men met back at camp around 4:00 that afternoon. Neither had bagged anything; in fact, Green had seen nothing. Monahan had a little better luck, spotting a big buck but missing his shot at it. Though disappointed at missing his chance at a trophy deer, he was nonetheless encouraged at finding game, itching for another crack at a trophy animal.

THURSDAY, AUGUST 29 The night was again miserably cold. Neither of the men slept well. Green dragged himself out at 4:30 and posted vigil on the nearby meadow. Monahan uncharacteristically slept on.

At sunrise Green returned to camp having seen nothing. The men ate breakfast and, as they had done the day before, headed off in different directions hunting for the elusive big game. Soon, the past two nights of poor sleep caught up with them. Overcome by fatigue and drowsiness, both men cut short their hunt and drifted back into camp shortly after noon. Neither man felt very well. That, plus the fact they were running low on food prompted their decision to return to the creek-bottom camp for medicine and more rations.

After a meager lunch the men gathered their gear and descended the steep mountain side to the canyon below. Along the way Bug cornered a porcupine which Monahan shot and kept. They also made mental note of the location of a recently used deer bed they discovered along the trail, thinking they might return and check on it over the next few days.

Back at their base camp they found everything in order and the horses looking well fed and content. With a quick supper of corn cakes, sorghum and tea, the exhausted men 'turned in' for the night.

FRIDAY, AUGUST 30 They awakened to rainfall, affording the men a good excuse for a much needed sleep-in. Happily, they awakened later to sunshine for breakfast and a good smoke.

It was not long before the sound of the nearby stream beckoned them with their fishing rods. They fished until 3:30 P.M., returning to camp with 50 fine trout and a refreshed outlook on life. Sunshine and fishing – the early 1900's equivalent to Prozac!

Little else is recorded for the day except that Monahan skinned out the porcupine they had collected on their return to camp the day before. The men planned to return to the mountain top the next day for another deer hunt. Monahan signed off his diary that day with a wistful "– providence smile on us!"

SATURDAY, AUGUST 31 It rained more in the night, clearing before Monahan got up that morning. Though apparently not quite freezing, it was cold enough that he built a big warming fire and started their breakfast of fish, coffee and sorghum with hot biscuits baked in the Dutch oven.

The weather looked promising for the day, so the men packed again for a two- or three-day hunt around the mountain top. They retraced their upward ascent arriving on top around 2:00 P.M. Cloud cover and warming temperatures threatened rain but it held off and the men were able to hunt as planned. Monahan stationed himself at the alpine meadow and brook, lying in wait of any feeding deer. Will Green took to the trail and hunted more broadly. At dusk both men returned to their temporary camp having seen no deer. Such persistence – most men would be drowning their sorrows at the Lardo Bar by then!

That night it rained again – heavily – and even though the men slept in the open air covered only by a spread tarp, they slept warm and dry. Bug was not so lucky as his lean-to shelter of fir boughs didn't shed water nearly as well.

SUNDAY, SEPTEMBER 1 The men got up in the rain and stoked a big fire. The chill of fall was in the air. The rain continued through the morning but the men decided to hunt anyway. They based their hope of hunting success in the rain on the prospect that the wet ground would provide good tracking and better stalking of deer.

Although the rain let up and the men hunted until about 4:00 in the afternoon, neither man found any deer – not even so much as a track. As they warmed and dried out by their campfire, Monahan and Green felt thoroughly discouraged and ready to call it quits. They concluded that deer were too scarce and shy in the region to make further forays worthwhile. Time was slipping away and they agreed to go elsewhere in their quest for deer. But before they could consider relocating their hunting efforts, they'd need to return to Lardo for more provisions.

MONDAY, SEPTEMBER 2 A night of restful sleep on the mountain top confirmed their decision. Will Green was up and had breakfast cooked and ready by the time his elder companion awoke. Oatmeal, coffee, corn cakes and bacon summoned Monahan awake in their alpine bivouac. Breakfast done, the men rolled up their scant trappings into two packs, departed their idyllic hideaway and slogged down the mountain again to the lower camp.

Arriving down at their Lake Fork River camp the men found their horses still faring well on the campsite's luxurious grass. Monahan was a little incensed when he discovered that their lone remaining stash of fruit, prunes, had been pilfered by marauding chipmunks: "...the infernal little rascals..."

Once re-settled back in camp, Green didn't wait long before lapsing into one of his favorite pastimes (drum-roll) – fishing! Of course, the fruits of this recreation, as usual, provided a fine dish for their supper.

TUESDAY, SEPTEMBER 3 The diary account of the following day notes a new affliction about to be visited upon Monahan. That and another notation poses a couple of natural history conundrums.

The day started routinely with Monahan setting out on a hunting quest. But this time he wasn't after the vexingly elusive mule deer. Instead, the expedition leader was in search of small game, "...a certain variety of chipmunk." There are no clues to the species he sought. My first inclination is that he may have been on a vindictive eradication of the "*prune-snitchensis*" variety, but I'm sure it was more scientifically based than that. Considering the ponderosa ("yellow pine") stands surrounding their permanent Payette Lake campsite, the Baker team probably had collected predominately the yellow-pine chipmunk (*Tamias amoenus*). Perhaps, in their present location – higher mountains with mixed coniferous woods – Monahan was in search of the slightly larger red-tailed chipmunk (*T. ruficaudus*), which has less affinity for ponderosa.

Even after reviewing Monahan's recently found field journals, I still cannot resolve this chipmunk conundrum . But I have a suspicion. None of Monahan's mammal records include technical (scientific) names – only common names, and even those are loosely applied (mouse, pine squirrel, little chipmunk). The expedition leader seems far less familiar with mammals than with birds. There are no clues to any literary reference or guide that he may have used to identify his mammal specimens. I doubt that his mammal sources would allow him to accurately discern between yellow-pine and red-tailed chipmunks, anyway. I suspect that during this trek into the high country Monahan may have encountered the over-sized chipmunk look-alike, the golden-mantled ground squirrel (*Spermophilus lateralis*). This squirrel is almost a dead ringer for a chipmunk, except much larger. Today, most campers and other outdoor enthusiasts still call this cute little food-begging rodent a chipmunk, or sometimes, because it resembles a chipmunk on steroids, "bullmunk". We'll revisit this question when it re-surfaces later.

An even greater mystery, to my mind, concerns what next befalls Monahan. With no further word on the outcome of his chipmunk hunt, Monahan returned to camp suffering from what is called "a dose of poison oak on his neck which he caught a few days before." Apparently, the rash had flared up and was becoming so bothersome to Monahan that the men decided to immediately break camp and pack the horses so they could return to Lardo to find some medicinal relief.

I'm surprised that Monahan would have encountered poison oak (*Rhus radicans*) anywhere in that region of Idaho. As one who is very susceptible to the toxic oils of this plant, I long ago learned to identify the shrub and am eternally on the lookout for it – for avoidance sake. In my wide meanderings about Long Valley and its surrounding terrain, including Lake Fork and Lick Creek areas, I've never observed it growing anywhere. Known interchangeably as poison oak and "poison ivy" its presence in Idaho seems largely restricted to lower elevation riparian areas.

Back on June 30, when the Baker expedition had descended from Centerville and Placerville down the grade to Horseshoe Bend, and back up out of the canyon toward Dry Buck, Monahan notes in his diary that he had a slight patch of poison ivy on his face. That terrain – lower elevations of the Horseshoe Bend foothills and the Payette River's drier bottomlands and brushy stream banks – certainly hosts scattered patches of poison oak. It is no surprise that the men encountered poison oak there. But now, two months later here in the high country of Lake Fork and Lick Creek, the plant's presence seems unlikely.

In any case, the rash was real and becoming increasingly miserable for Monahan. By 2:30 that afternoon the men were on their way to Lardo.

By dusk the men were back at their Payette Lake camp. Supper was quick in the making and within the hour the weary men were in bed. Rash not withstanding, the men slept well that night.

Large-flowered Collomia

· 13 ·

REST AND RECUPERATION AT PAYETTE LAKE

SEPTEMBER 4 - SEPTEMBER 13

WEDNESDAY, SEPTEMBER 4 Monahan and Green were up early. Right after breakfast the men visited the Gaekels, probably to reclaim their stowed camping gear. Learning of Monahan's poison oak rash, Mrs. Gaekle graciously provided some "plantain lions" for making a compress to be applied to his rapidly spreading rash. Herbalists have long known that crushed leaves of some species of plantain contain certain astringents that can be used to soothe sores, bites, and the rash of poison oak and poison Ivy (Vizgirdas 2003). Although Idaho has two introduced species of plantain, *Plantago lanceolata* and *Plantago major*, both regarded today as "weeds", the plants are edible and were likely grown as a food source in early homestead gardens.

The rest of the day was spent in camp periodically treating Monahan's rash with the medicinal plantain and reading the two dozen papers sent by a friend back in Kansas. The irony of this vexing skin rash is this: two decades later – midway through his medical career - Monahan would spend a year in Vienna, Austria studying skin diseases to begin his specialization in dermatology.

For the first time since August 3, the field journals again record the taking of specimens, neither of which is mentioned in the diary account. Big game hunting and Monahan's recuperation from the poison oak rash had effectively cast a temporary ceasefire on small animal collecting.

~~~~~~~~~~~~~~~~~~~~~

(Birds) *9/4 No. 100 Pandion haliaetus (Osprey)*
Payette Lake Boise Co. Idaho (rare)
L of body 8.5 depth of front 5 width 4.5 depth of hind 4 width 3.75
L of neck 4

(Mammals) *9/4 No. 21* [Tamiasciurus hudsonicus] *(Pine Squirrel)* [Red Squirrel]

THURSDAY, SEPTEMBER 5 By now Monahan's poison oak rash was responding favorably to the plantain treatments. He felt well enough to venture out on a short but unsuccessful junket in search of new birds. The apparent absence of heightened bird activity led Monahan to conclude that fall migration was not yet underway. Nonetheless, he was well aware of the signs and feel of autumn and knew that soon winter and deep snow would settle into Long Valley; he reflects on the fact that he hopes to be long gone to warmer climes by then.

In keeping with the advent of fall weather it had rained again that day, and continued throughout the night.

A lone specimen, unnoted in the diary, is recorded for this date in a field journal.

(Mammals) *9/5 No. 16* [Lepus americanus] *(Snowshoe Rabbit)* [snowshoe hare]
Taken in Boise Co. Ida
l of body 12 in neck 2.75 depth of front body 4.5 flank 3 eyes brown
~~~~~~~~~~~~~~~~~~~~~

FRIDAY SEPTEMBER 6 The day broke cool and cloudy. The ominous threat of a storm kept Monahan and Green close to camp puttering at small chores. The diary notes that they worked on their boots, getting them ready for their next trek into the mountains after big game. Just as soon as Monahan's rash healed sufficiently they planned one last hunt for mule deer in the high country.

SATURDAY, SEPTEMBER 7 Antsy for activity the Baker men took another short outing. The Gaekel's son, Walter, led them up 'Bowyer Trail' to a place they called "Crater Lake". For the first time in a long while, the diary notes the collection of some specimens. First, it notes that they shot "...two fine jays". These are not specifically named, nor do the field journals record any corresponding entries, so we don't know if they were Steller's or gray jays. The other collection report was of "...some fine dragon flies." Monahan's diary

contains so few accounts of collected insects, that this one stands out, particularly because it names the type of insects collected – though not by species. It is unlikely that Monahan possessed any Idaho entomological source – or if there was even one available in 1907 – that could identify the dragonflies found at Idaho's mountain lakes. Only in recent years have useful and accurate field guides to dragonflies become available. From my own limited collecting and observations around Payette and Upper Payette lakes, Monahan specimens may have included colorful examples of: the variable darner (*Aeshna interrupta*), mountain emerald (*Somatochlora semicirularis*), white-faced meadowhawk (*Sympetrum reubicundulum*) and twelve-spotted skimmer (*Libellula pulchells*), among other species.

Depending upon how loosely Monahan classified "dragonflies" he may also have included damselflies like the familiar bluet (*Enallagma civile*) and common spreadwing (*Lestes disjunctus*).

Curiously, the "two fine jays" never made their way into the field journal records as collected specimens. Perhaps they were trumped by the other, highly unusual specimen collected during the hike that day. It was a lone, abandoned domestic sheep. The law of the land at that time was that any lost sheep discovered to be left behind in the mountains was "fair game" for the taking. Separated from the band and left behind boded certain death – eventually by winter if not sooner by predators. Therefore, it was considered an act of mercy for the sheep as well as a wind-fall gift for the finder. Especially finders who had twice-over eaten their weight in fish!

After dispatching the animal they field dressed it and left it covered with fir boughs until their return trip down the trail. To ease the load of packing it home they ran a long pole through the length of it and carried it between them. They light-heartedly dubbed it their "deer", and Monahan penned diary comments on the comical appearance of the entourage, contending that for any onlooker it would have been worth the price of admission to see them trooping along. The sheep must have fed well on its mountain fare for it was quite heavy and the merry hunters soon wished for some other means of conveyance, be it a horse, mule, or auto. But after many rest stops they finally made their way to the Gaekel farm.

There they found Mrs. Gaekel milking cows as her husband was away for a time, and she insisted that the Baker men stay for supper, which they did, gladly.

SUNDAY, SEPTEMBER 8 The weather had warmed and remained dry. Another Sunday passed quietly in camp. Other than reading, smoking and sewing a little the only noteworthy event seemed to be when Green prepared their first serving of mutton roast. Due to some cooking error, it burned to a crisp – and they ended up having bacon instead.

Without photos from the expedition and with scant physical descriptions of either Will Green or Monahan, today's diary entry offers a hint of an answer to one question that I've pondered: did either of the men grow a beard over the summer, or were they more or less clean shaven? Back on June 30, the diary records that the men gave each other haircuts. There are two or three other diary notations about their taking baths. But there is nary a word anywhere in the diary about shaving. However, this day's report of Monahan's healing progress provides us a rare snapshot. "Cot's (Monahan's) poison is about healed but he looks like a red onion half peeled with his short hair long beard (2 weeks old) and sunburned nose."

His beard was only two weeks in the growing. And that growth probably resulted from the bivouacking in the high country over the past 10 days, followed by his rash attack. Obviously, Monahan didn't sport a summer's growth of beard. As to Will Green, I find no clues one way or the other.

MONDAY, SEPTEMBER 9 Their mutton breakfast that morning was far more successful than Green's burnt offering the previous evening. Monahan notes: "fine old meat fried in bacon grease and flour."

The next two days were pleasantly cool and the men did little more than sleep in and relax around camp. This prolonged period of rest and recuperation was not only allowing Monahan's rash to heal, but also gave the men time to catch-up on some routine tasks like washing and mending clothes, and repairing boots.

Two bird collections are recorded only in the field journals at this time.

~~~~~~~~~~~~~~~~~~~~~

(Birds) *9/9 No. 98 Pileated Woodpecker [Dryocopus pileatus]*
Payette Lake Boise Co. Ida.

9/9 No. 99 Sitta canadensis (Red-bellied [Red-breasted] *Nuthatch)*
rare

~~~~~~~~~~~~~~~~~~~~~

TUESDAY, SEPTEMBER 10 The forced idleness was beginning to wear on the men. They were restless to head back into the high country in search of the elusive deer. As Monahan wrote in his diary, "Getting anxious to get back to the 'fruitless hunt' & try try again..."

A lone bird specimen is recorded for that date in the field journals.

> (Birds) *9/10 No. 101 Asyndesmus torquatus [Melenerpes lewis] (Lewis['s] Woodpecker)*
> *Payette Lake Boise Co. Ida*

WEDNESDAY, SEPTEMBER 11 Green helped the Gaekels in the hay, which probably helped pass the time for him. The valley was enjoying a stint of fair, cool weather and harvest time was in the air.

Left alone in camp, Monahan occupied himself by going to town and checking on the mail. Otherwise, the recuperating expedition leader "...ate 'alone in his tent' like the proverbial Arab."

That night Monahan and Green visited with the Gaekels and their guest, Fred Williams, a newly arrived valley resident from Pennsylvania. Williams was a bachelor and a rank Socialist and the Baker men enjoyed a lively discussion of politics, visiting late into the evening. When it was finally time to return to their camp, the Baker men borrowed a lantern to find their way. Monahan's diary entry provides one of the best descriptions of their camp location along the peninsula: "...had to borrow Walter's lantern to guide us through the great pines along the lake to our camp – almost a mile up the shore."

The following day Green again worked in the hay with the Gaekels and "...got a square meal", suggesting the bartering of food for his labors: work for a home-cooked meal, a true luxury for the long-time campers.

Monahan, meanwhile, collected some birds along the lakeshore and spent the day working on their skins. The diary reveals no clue to the identity of these particular birds. And here again, the recently exhumed field journals don't resolve this matter; they record nothing on this particular date.

That evening the Baker men again stayed after supper and talked "Socialism" politics. It must have been another lively and productive session because Monahan and Green didn't get to bed until 11:00 that night.

FRIDAY, SEPTEMBER 13 Though it had threatened rain the evening before, dawn broke as another one of those beautiful fall days known in Idaho as "Indian summer".

Green worked haying one last day. While he labored on the nearby Gaekel farm, Monahan passed the time as productively as he could. First he went into town to check for mail. Another part of the day was spent hunting for the osprey that made occasional visits to the lake. And again, the diary records that Monahan spent some time taking pictures. And again, this author wishes, so longingly, for any copies of them. Any photos from that 1907 trek would be priceless.

Shooting Star

·14·

THE BEGINNING OF THE END?

(SEPTEMBER 14–NOVEMBER 1)

SATURDAY, SEPTEMBER 14 At long last, the day arrived for the Baker team to pack up and head for the high country. Monahan was sufficiently healed from his rash to resume their expedition work. It was time to make one last push to collect some deer, bear, and mountain goats for the Baker University Museum. A mood of both anticipation and cautious optimism hung lightly over the men as they prepared for departure.

At 9:00 A.M. they set out on foot, each man leading a heavily packed horse. They were bound for their favorite campsite at the rapids on the Lake Fork River, arriving around 3:00 P.M.

By the time they unloaded their horses and set up camp they were fairly famished, so they quickly prepared an early supper. On the hike over that day, Jo (a.k.a. Bug or Bughead), who was now beginning to show real promise as a hunting dog, had pointed a ruffed grouse which Monahan shot. That not only filled his master's heart with pride, but filled the supper pot as well.

After supper, Green fished while Monahan went out hunting for chipmunks and rock rabbits. I suspect the targeted chipmunk was the same "certain variety" – either the red-tailed, or even more likely, the chipmunk look-alike, the large golden mantled ground squirrel – that Monahan hunted near here 11 days earlier (the day poison oak suddenly flared). This is the first mention of rock rabbits, or pikas, even though much of the men's high country deer and goat hunting among the rocky screes was in prime habitat for those diminutive rabbit cousins. Of course at that time the hunters were armed with heavy caliber rifles which would have blown the little mammal to bits, in addition to alerting any big game in the vicinity. Monahan probably wanted to collect the chipmunks and pikas with lighter caliber weapons before they started after the big game with heavy artillery.

Neither the hunter nor the angler were successful.

SUNDAY, SEPTEMBER 15 Today was the official opening day of the big game hunting season. Though a Sunday, it was a long travel day for the Baker team. And contrary to the spirit of Sabbath rest, the day imposed perhaps the most grueling strain on the horses of the entire summer. Their destination was high above Lake Fork to a sheep camp run by a Mr. Chapman. The attraction of this particular site was that the sheepherders there were reporting sightings of bear and mountain goats. Unfortunately, getting to the sheep camp meant traversing a vast expanse of old forest burn. The downed logs strewn across the trail were innumerable and often insurmountable. Time after time the horses had to jump over or go around fallen trees. Monahan describes some of those logs as hundreds of feet long and as much as ten feet thick. At one point old "Star Crust" in attempting to leap a log failed to clear the barrier and fell back on his pack, rolling over and over into a gully. Miraculously, the old horse was not injured and very little damage was done to the pack contents. The one notable casualty of that accident was the men's coffee pot. Surely, that had little to do with the fact that the Monahan diary comes to an end two days later, even though the Baker team remained in Idaho until November 1.

The men eventually reached the sheep camp and wearily pitched their own camp nearby. After lunch they settled their horses into a good grassy spot, then visited the boys at the sheep camp. The herders were expecting them and insisted they stay and have supper with them. Fresh mutton, tea and nice biscuits were too enticing to turn down and the Baker men accepted gratefully, even though they had just eaten their late lunch not long before.

On their way back to their camp that evening Monahan and Green finally enjoyed a turn of good luck in the quest for big game animals – they spotted and shot a bear. It was small and of a black and brown coloration; both a little disappointing to Monahan. But he seemed grateful for such as it was.

This bear specimen doesn't show up anywhere in the field journal records. The only thing recorded for this date is another Williamson's sapsucker, the fourth collected thus far. Note that the location given is Payette Lake, which they are nowhere near on this date. Thus continue the unexplained discrepancies between diary and field journal accounts. Still, I adhere to my practice of entering the field journal records on their stated dates, rather than in the sequence of their numbers.

(Birds) *9/15 No.114 Sphyropicus [Sphyrapicus] thyroideus (Red-throated Woodpecker)* [Williamson's Sapsucker]
Taken in woods on Payette Lake Boise Co. Idaho- rare

MONDAY, SEPTEMBER 16 After skinning out their bear kill of the evening before, Monahan and Green set out on a long hunting tramp through the area. It was as fruitless and discouraging as all the others the two men had endured in their summer-long quest for big game. One detects a profoundly disheartened note in Monahan's closing of his diary entry that day: "Took a long tramp but no results – same old story – nothing doing – fiddle on such a country. Anyway we are both tired of the luck we are having." Was this the death knell to the 1907 diary?

After striking out once more on big game, the men must have settled for some smaller specimens. These two dippers are recorded in the field journals on this date – the only ones collected all summer.

(Birds) *9/16 No. 116 Cinclus mexicanus (American Dipper)*

9/16 No. 117 Cinclus mexicanus (American Dipper)

No location or other supportive data accompany these journal entries. But at this time we know that the Baker team was roaming mountain landscapes carved and fed by tributaries of the Lake Fork River – suitable habitat for this uniquely aquatic songbird. Found only in the western U.S., this species was likely a novelty to Monahan and Green.

162 **TUESDAY, SEPTEMBER 17** Today's diary entry is a mystery! It begins as usual with the date, but then there is nothing . . .only blank space. To end the long, unbroken string of 88 daily entries – the expedition's events, observations, musings and wanderings – there comes a striking void of words. The diary of the 1907 expedition abruptly and eerily slams shut.

Why? Why this unexpected halt to the diary's narrative, even though an end-note states that the men remained in Idaho until November 1st ? Had Monahan and Green grown so despondent about their woeful lack of hunting success that they couldn't bear reporting more failure? If that was the case, say… over the next few days, everyone knows the difficulty of going back to fill in the blanks of fading memories. Often, after such a hiatus, fading details and inaccuracy of daily events and notes make any attempt to restore the continuity very difficult. Could that have been the case with Monahan and Green? There is nothing to help explain this unexpected end to the diary.

The diary's end-note entry solemnly pronounces its premature demise. "And now diaries like all things earthly, come to an end and sometimes untimely – so it is with this one."

It continues in a terse summary of the conclusion to the 1907 Baker expedition. I think it is worth sharing:

"We left Idaho Nov.1. We succeeded in getting 4 nice deer to send in."

Besides the four deer, what other specimens were collected by Monahan and Green between September 17 and November 1? With fall migration of birds in progress during that time, it would be difficult to believe that there was not considerable collecting still going on during those last unaccounted for six weeks in Idaho. Yet, the diary is mute.

What transpired during those concluding six weeks in Idaho would have remained a huge mystery had it not been for the enlightening information found recorded in the recently-acquired Monahan 1907 field journals. The collecting notes in his journals continue on more than two weeks beyond the silenced diary. They are listed below in chronological order of their listed date, despite some irregularities of their number sequence.

~~~~~~~~~~~~~~~~~~~~~

(Birds) *9/17 No. 106 Cyanocitta stelleri (Steller's Jay)*
Common

9/18 No. 102 Sphyropicus [Sphyrapicus] thyroideus (Red-throated Woodpecker)[Williamson's Sapsucker]

9/18 No. 107 Cyanocitta stelleri (Steller's Jay)

9/18 No. 108 Cyanocitta stelleri (Steller's Jay)

9/18 No. 109 Sitta canadensis (Red-bellied [Red-breasted] *Nuthatch*
(rare) Payette Lake Boise Co. Ida
~~~~~~~~~~~~~~~~~~~~~

9/18 No. 110 Junco oregonus (Oregon [Dark-eyed] *Junco)*
Common

9/18 No. 111 Larus californicus (?)(Coues) (California Gull) juv.

The identification of this bird is not at all clear & until other authorities can be read I am not ready to say it is L. californicus as it bears characteristics of several others of the L. group.

Taken on Payette Lake Boise Co. Idaho, solitary.

L. of body 5 in Neck 4.5 girth of body 7.5

The above specimen is Monahan's only recorded encounter with any gulls in Idaho. Since it was apparently a juvenile bird, it presented a bit of an identification conundrum. The California gull is one of the two more common gull species frequenting the lakes and river courses of southwestern Idaho, so Monahan probably had it right.

9/20 No. 103 Picicorvus columbianus [Nucifraga columbiana] (Clark's Nutcracker)

Lake Payette Boise Co. Ida

9/20 No. 104 Picicorvus columbianus [Nucifraga columbiana] (Clark's Nutcracker)

(as above)

9/20 No. 115 Picicorvus columbianus [Nucifraga columbiana] (Clark's Crow or Clark's Jay) [Nutcracker]

There's a curious shift in common name usage for the Clark's Nutcracker on this last entry (no. 115 above) - this despite the stated name "Clark's Nutcracker" on both of the previous two specimens. Yet, Coues (1903) actually lists Clark's Crow as the first choice of common names for the bird, followed by Clark's Nutcracker.

~~~~~~~~~~~~~~~~~~~~~

9/20 No. 118 Sitta canadensis (Red-bellied [Red-breasted] *Nuthatch*
Notes as in # 109

9/20 No. 119 Sitta canadensis (Red-bellied [Red-breasted] *Nuthatch*
Notes as above

9/22 No. 112 Bonasa umbella [umbellus] (Ruffed Grouse)
Lowland, Lake Payette Boise Co. Idaho Common

No. 113 Bonasa umbella [umbellus] (Ruffed Grouse)
Notes as in # 112

~~~~~~~~~~~~~~~~~~~~~

Even though the date is lacking for No. 113 above, I've listed it on the same date as No. 112 because it follows in sequence in the field journal and is more or less implied to have been collected along with No. 112.

~~~~~~~~~~~~~~~~~~~~~

9/22 No. 120 Pica pica hudsoni ([Black-billed] *Magpie)*

9/22 No. 121 Cyanocitta stelleri (Steller's Jay)
~~~~~~~~~~~~~~~~~~~~~

9/24 No. 122 Bonasa umbellus (Ruffed Grouse)
Notes as in 112

Here begin the entries of the last seven Idaho bird specimens recorded for the 1907 summer expedition. The first, No. 123, is the only one bearing a date, October 2. I've listed the remaining six in their sequential order as listed in the field journal, even though they bear no date with them. Uncertain of their exact date of collection, I assume they too were added sometime later in October, if not on October 2.

---

166 *10/2 No. 123 Ceophloeus [Dryocopus] pileatus (Pileated Woodpecker*
Payette Lake Boise Co. Ida

No. 124 Canace [Falcipennis] canadensis ("Fool Hen") Spruce Grouse

No. 125 Canace [Falcipennis] canadensis ("Fool Hen)

No. 126 Canace [Falcipennis] canadensis ("Fool Hen")

No. 127 Canace [Falcipennis] canadensis ("Fool Hen")

No. 128 Canace [Falcipennis] canadensis ("Fool Hen")

The above usage of *Canace canadesis* for the spruce grouse puzzled me at first. Both of the turn-of-the-century bird references – Reed and Coues – which Monahan cites in his field notes (and thus, must have carried with him) list *Canachites canadensis* as the current, accepted scientific name of the spruce grouse (fool hen). However, a detailed reading of the species account in Coues' final (1903) edition also finds the statement that *Canace canadensis* was the accepted name listed for the species in former editions of his *Key to North American Birds.* This not only explains the technical name confusion, but also suggests that Monahan's expedition copy of Coues' must have been one of the older, pre-1903 editions.

The final bird specimen listed in his journals is a juvenile common loon. As with the previous five grouse specimens, there is no date listed beside its entry. And, like the grouse, it's unclear whether this loon was collected on October 2, the last date listed, or on some later date.

~~~~~~~~~~~~~~~~~~~~

No. 129 Gavia imber [immer] (Loon) [Common Loon] *Juv* 167

L 32 Ex 48

L of body 12:00

L of neck 7:00

g of front 12:00

" of mid 11:50

" hind 9.50

Remark: Back lines straight; belly gentle curve, whole body tapering to point Legs situated 6 in from front & just after middle up & down; Eyes brown. Shot from shore early in the morning while it was feeding on the algae in shallow water.

~~~~~~~~~~~~~~~~~~~~

Thus ends the entries of Idaho bird specimens. Further specimens are listed in the journal, but they don't begin again until November 20, with No. 130 listed as a "Valley Quail" collected in Josephine Co., Oregon.

This last Idaho bird journal entry piques the curiosity of an ornithologist. For a fish-eating loon to be described as "feeding on algae" seems odd. Perhaps, Monahan simply meant that the bird was foraging *amid* the algae growth of the lake?

Monahan's mammal journal likewise records the last of the Idaho mammal specimens. The last entries confirm the diary's concluding remarks about the successful collection of four deer. As with the diary reference to these deer, no dates accompany their record. The deer entries simply follow the last dated collection of the two red squirrels on September 9.

~~~~~~~~~~~~~~~~~~~~~

(Mammals) *No. 23 (1)* Deer *Fawn*

L hip to shoulder	*24*
Girth front	*28*
" hind	*19*
Depth front	*19*
" hind	*19*
Sh[oulder] *to knee*	*19*
Hip " "	*19*
Neck	
Front girth	*10*
Mid "	*11*
Hind "	*13*
Length	*8*

Eyes
Ebony pupil
Color brown

No. 24 (2) Deer Buck

L hip to shoulder	*31*
g[irth] front	*38*
" hind	*30.50*
~~~~~~~~~~~~~~~~~~~~~

Depth front 25
" hind 25

Neck
L 15
girth front 15
" mid 16
" hind 19

No. 25 (3) *Deer Fawn*
L of hip to shoulder 25
W of front 20
" " hind 20

Neck
L 10
g front 12
" mid 14
" hind 16
girth of front 28
" hind 22

No. 26 (4) *Deer Doe*
Dist hip to shoulder 30
g front 36
" hind 30
D front 24
" hind 24

Neck
L 9
girth front 14
" mid 15
" hind 18

~~~~~~~~~~~~~~~~~~~~~
~~~~~~~~~~~~~~~~~~~~~

Like the bird collection journal, the next series of entries in the mammal journal begins on November 20, 1907 in Josephine Co., Oregon as the location.

There is not a word on how or where the Baker team exited Long Valley. The succinct conclusion to the diary account jumps from their fall hunting at Payette Lake to the disposal of their wagon and horses in the Snake River. We are left to fill in the course of events between.

The diary description of the disposal of their team of horses and the old wagon into the Snake River offers a most poignant end to the 1907 expedition.

"We found it necessary to shoot our poor horses as they were just about past going and no one would take them on the account of the scarcity of hay and grain. We led them gently away to a place where the Snake River made a quiet bend and while they were still happy, one at a time fell to rise no more. The winding current of the old river carried them from our view and old Bally and Star Crust were no more. We pushed the old wagon after them and made our way back to camp with heavy hearts." (Socolofsky 1971).

This disposal of their team and wagon poses yet another big question: where on the Snake River did that occur? If, as Socolofsky reports in the published diary, the Baker team was driven from Idaho in late October by severe snows, what would have been the most prudent departure route from Long Valley to the Snake River? I'm guessing by way of Meadows and on to Weiser.

Several advantages can be surmised for that route. First, with fall weather dropping snow in the high country, the lower elevation of the Meadows-Weiser route had considerable travel advantage. A second advantage is that from Long Valley to the Snake River it would be a considerably shorter route to Weiser than returning over the same route the expedition traveled to get to Payette Lakes in the early summer. We must also take into consideration some implicit logistics facing Monahan and Green regarding the handling and deposition of the expedition's summer payload of specimens.

If they disposed of their wagon what did they do with their sizable accumulation of Idaho specimens? It isn't likely that they would freight them along to Oregon. The end-note mentions the "4 nice deer to *send in*"; most likely along with their whole summer's collection, before traveling on west.

The Weiser route satisfies several needs of the Baker team. Beyond just a quick, lowland route out of the mountains, the Pacific and Idaho Northern Railway system provided a means of exporting their summer's cache of museum specimens back to Kansas and Baker University. If packing materials were not available at Evergreen, Weiser no doubt provided the necessary items with which Monahan could box, crate and otherwise prepare the specimens for safe shipment back to Kansas. That would secure the safe delivery of their summer's holdings and allow the men to journey unencumbered and lighter on to Oregon for the next three months' segment of the expedition. Lastly, not far from Weiser flows the Snake River, the ultimate repository of their serviceable but no-longer needed horses and 'Science Car'.

Evening Primrose

·15·

EXODUS TO OREGON

(NOVEMBER 1, 1907–FEBRUARY 10, 1908)

After the Baker team left Idaho on November 1, the mystery of their exact whereabouts and activity persists. Beyond "Josephine County, Oregon" there is no given destination for the two men, assuming that they remained together during the Oregon junket. The field journals' listings below offer the best clues to the dates and duration of the Oregon stay. Whereas the Idaho diary succinctly states that the team left Idaho on November 1, the Oregon specimen records begin on November 20, 1907 and conclude on February 12, 1908. In the following chapter (Chapter 15 – Monahan the Man) I discuss a shred of evidence stemming from Monahan's early childhood that might suggest the possibility that he home-based in Grant Pass, Josephine Co., Oregon, during this stint of specimen collecting.

~~~~~~~~~~~~~~~~~~~~~

(Birds) *11/20 No. 130* [*Mt. Quail* is crossed out]
"Valley" [California] *Quail*
Josephine Co., Ore.

11/20 No. 131 [*Mt.* is crossed out] *"Valley" Quail"*
[California] *Josephine Co.*
~~~~~~~~~~~~~~~~~~~~~

11/20 No. 132 Melanerpes formicivorus bairdii (Calif. W. pecker) [Acorn Woodpecker]
Co. Ore

No. 133 M. formicivorus bardii (Calif. W. pecker)[Acorn Woodpecker]
As in # 132
L of h 2.75 W of h 2 d of h 1.50

11/14 No. 134 Colaptes cafer Red-shafted W. pecker* [Northern Flicker]
Josephine Co. Ore
L of h 3.75 W of h 3.00 D of h 2.75

*Date out of sequence, but placed here in correct order.

11/28 No. 135 Merula migratoria propinqua [Turdus migratorius] Western Robin [American Robin]
Josephine Co. Ore

1/1/1908 No. 136 Quail [Presumably "Valley" or California Quail]

1/1 No. 137 Colaptes cafer Red-shafted Woodpecker [Northern Flicker]
Josephine Co. Ore

1/4 No. 138 Junco Oregonus [hyemalis; Dark-eyed Junco, *oregonus* form]
Josephine Co. Ore

1/4 No. 139 Junco Oregonus [*hyemalis;*Dark-eyed Junco, *oregonus form*
Josephine Co. Ore

1/4 No. 140 Colaptes cafer [Northern Flicker]
Josephine Co. Ore

1/4 No. 141 Colaptes cafer [Northern Flicker]
Josephine Co. Ore

1/4 No. 142 Junco Oregonus [hyemalis; Dark-eyed Junco, oregonus form]
Josephine Co Ore

1/6 No. 143 Junco Oregonus [*hyemalis;* Dark-eyed Junco, oregonus form]

1/10 No. 144 Colaptes cafer [Northern Flicker]
Josephine Co. Ore

No. 145 Colaptes cafer [Northern Flicker]
Josephine Co. Ore

No.146 Sphyropicus varius ruber [Sphyrapicus ruber] Red-breasted Woodpecker
[Red-breasted Sapsucker]

2/10 No. 147 Aphelocoma floridana woodhousii [*Aphelocoma californica;* Western Scrub-jay]
Josephine Co. Ore

2/10 No. 148 Aphelocoma f. w. [*Aphelocoma californica;* Western Scrub-jay]
As above

2/10 No. 149 Aphelocoma f. w. [*Aphelocoma californica;* Western Scrub-jay]
Josephine Co. Ore

2/10 No. 150 Sphyropicus varius ruber [Sphyrapicus ruber]
Josephine Co. Ore

2/10 No. 151 Chroicocephalus philadelpha [Larus philadelphia; Bonaparte's Gull]
Josephine Co. Ore

2/10 No. 152 Aphelocoma f. woodhouii [Aphelocoma californica; Western Scrub-jay]
Josephine Co. Ore

2/10 No. 153 Aphelocoma f. woodhousii [Aphelocoma californica; Western Scrub-jay]
Josephine Co. Ore

~~~~~~~~~~~~~~~~~~~~~

175

It is here that Monahan's long lost 1907-08 avian field journal concludes with these 24 Oregon bird specimens. Of the eight species produced by this sodden Oregon extension of their Idaho trek, only five were new to their collection: California quail, acorn woodpecker, red-breasted sapsucker, western scrub-jay and Boneparte's gull.

Like the Idaho portion of their expedition, the Oregon foray collected few mammals. The following Oregon entries conclude Monahan's 1907-08 mammal field journal.

~~~~~~~~~~~~~~~~~~~~~

11/20 No. 27 [Spermophilus beecheyi]"Digger Squirrel" [California ground squirrel]
Josephine Co. Ore
L of b 7 G f 6, G m 7 G h 5 Hip 2

No. 28 [Spermophilus beecheyi]"Digger Squirrel"[California ground squirrel]
As in # 27

11/25 No. 29 Sciurus fasson [*Sciurus griseus*] *Silver-gray Squirrel* [Western Gray Squirrel]
Josephine Co. Ore
L of b 9 G f 8 G h 7 hp 3 Length 28.5

1/1/ 1908 No. 30 [northern] *Flying Squirrel* [*Glaucomys sabrinus*]
Josephine Co. Ore

1/2 1908 No. 31 [Dusky-footed?; Bushy-tailed?] *Wood Rat [Neotoma sp?]*
Josephine Co. Ore

No. 32 [Black-tailed] *Jack Rabbit* [jackrabbit] [*Lepus californicus*]
Josephine Co. Ore

1/4 1908 No.35 Panther* [puma] *[Puma concolor]* – *Male adult*

Measurements –

Length	*51 in*
Sh to hp	*29 "*
Hp " knee	*14 "*
Sh " "	*15"*
Body	
Girth front	*32 "*
" Mid	*30"*
Neck	*26"*
Length	*7 "*
Girth – hind	*7"*
" Mid	*5 ½ "*
" f	*5 ¾"*
Back on Tip W	*6 ½*

* Presumably taken in Oregon with the above specimens, but the field journal record is unclear.

~~~~~~~~~~~~~~~~~~~~~~
~~~~~~~~~~~~~~~~~~~~~~

PART III

RESTORED TO MEMORY

·16·

THE SUPPORTING CAST

With this book's illuminating disclosure of mislaid records of Monahan's 1907 expedition, clearly it was a grandly successful mission. Returning with 153 bird specimens, 37 mammals, 18 sets of eggs and all manner of yet unknown botany, insect and herptile specimens, Baker University's Museum of Natural History reaped its bounty. Given our country's extremely fragile economic times, and the trek's remote, rugged destination, the feat is all the more remarkable.

Though Monahan was the apparent mastermind and driving force behind this venture, he enjoyed the unflagging and able assistance of his young expedition partner, Will Green. From all indications the two maintained a genial and smooth working relationship.

I wonder what factors played into Green's enlistment into this nine-month-long, demanding enterprise? Had this newly minted Baker graduate been a promising student, tutored and groomed by Monahan in some of his classes prior to the expedition? Were they co-conspirators of the expedition from its inception? Certainly, the expedition would have been impossible to accomplish single-handedly. The fact that Will was little brother to Monahan's close friend and former classmate, Raymond Green, may also have held some sway in his joining this expedition.

What we do know about Will Green is that after returning to Kansas from the 1907 expedition he settled in Hutchinson where he worked as a fire insurance and mortgage agent before moving to Winnipeg, Canada in 1910. For over a decade he worked there in the Grain Exchange. Census records show that two sons were born to him there: William D. Green Jr. in 1917, and Stewart, in 1918. Apparently, the lingering influence of his 1907 memories finally prompted his move back out west in 1921 where he set up his real estate and insurance business in Portland, Oregon. It is interesting that both Monahan and Will Green had first born sons that carried their fathers' given name. Will Green Sr. lived out his life in Portland dying in 1949 at age 64 (Socolofsky 1971).

At the time Ray Green joined Monahan and his brother that summer of 1907, he was a school administrator in the Concordia, Kansas High School. Soon thereafter he bought into the Concordia newspaper, the *Daily Blade,* ultimately taking over the business. His zest for fishing that was so readily apparent during his 1907 adventure with the Baker team never abated. He continued his outdoor pursuits of fishing and hunting, even occasionally traveling out west to Idaho and the Rogue River country of Oregon. He outlived his brother Will by a year, passing away in 1950 at age 67.

The one native Idahoan in this story, Thomas Clyde Bacon, returned to his home state upon graduating from Baker, and remained there. The Bacon Family must have had Methodist ties in Boise. Why else would their son have ventured so far away from Idaho, specifically to Baker University? In 1907 the presiding pastor in the Boise First Methodist Church, Dr. Dalton, was a Baker alumnus, as was his predecessor, Dr. I.F. Roach. In all likelihood, they actively recruited students for their alma mater, especially from among their parishioners.

I still strongly suspect that the Bacon family bore some influence on the Baker team's choice of Idaho as their expedition destination. Similarly, I don't think it was any accident that Monahan and the Greens postponed their hunting of big game in the high country until Bacon joined them late in that summer of 1907. I wager the three Kansans regarded him as their "native" hunter; the one most familiar with Idaho big game and their likely haunts.

In any case, his Idaho roots retained a strong pull on Bacon. After his return to Idaho he remained in Boise until 1912, then relocated to Twin Falls. Throughout the rest of his long life – he died one year after Monahan, in 1965, at age 80 – Bacon established quite a name for himself in the Idaho and National Woolgrowers Association. Additionally, he served broadly in credit associations, community politics and area charitable organizations (Socolofsky 1971).

In the published diary account, Socolofsky (1971) contends that the four Baker men of the 1907 expedition maintained their friendship throughout the remainder of their lives, but Monahan Jr. recalls little of his dad's continued contact with any of them, except that he spoke often of Will Green. Whatever their connection through the subsequent years, no doubt their lingering memories of that 1907 adventure in the rugged Idaho mountains remained a lasting bond between them.

Lupine

·17·

MONAHAN, THE MAN

Elmer Perry Monahan was born to John and Bessie (O'Connell) Monahan in Darlington, Wisconsin on June 8, 1874. John's family background remains uncertain but we know that Bessie was born in New York City. Elmer was third in the line of seven siblings, the second of four boys. Surely, supporting their steadily growing family kept the Monahan's strapped and exploring all available resources. Early on, the family made their way to Nebraska.

As was fairly common with large families in those days, young Elmer spent some time living with other families in Syracuse and Unadilla, Nebraska. One of those foster families, the Hair family, boasted of kinfolk out west in Grants Pass, Oregon who were prominent proprietors of the locally known "Big Hair Hardware Store". Let me interject a crucial point of interest here. This smidgen of Hair family history may figure largely in solving one of the yet unresolved mysteries of the 1907 expedition. That is, *where* in Josephine County, Oregon did Monahan and Green home-base after leaving Idaho in November? Neither the diary, nor the field journals divulge the Baker Team's exact whereabouts in Oregon, beyond just "Josephine County". Of course, a corollary question here is, *why* did Monahan and Green end up in Josephine County, of all the biologically rich areas of Oregon? Coincidentally, Grants Pass, Oregon, home of the Hair Family's Big Hair Hardware Store is nestled within Josephine County. Could it be that Monahan later reconnected with his early childhood family friends, the Hair family? I have no proof of this but in the absence of any other explanation that connection seems plausible.

At about age five, Elmer was taken in permanently by a kindly family in Baldwin City, Kansas. This arrangement with the Clyde and Beula Stewart family seems to have been a very favorable and nurturing setting for Elmer. Their continued influence clearly paved the way for young Monahan to grow strong and confident on his long, positive pathway to a successful adulthood. Whether or not they were well educated themselves, the Stewarts must have placed importance on education. They enrolled Elmer in high school at the Baker University Academy, a private school associated with the Methodist college. This close connection to Baker University no doubt fostered his eventual matriculation at that university following high school.

I know nothing about the fate of the rest of the Monahan family except that Elmer was the only member of his branch of the family to attend college, and that his parents are both buried in Palmyra, Nebraska. Despite his early childhood uprootings and transplantings, it seems that Providence smiled upon the upbringing of young Elmer Perry Monahan.

There is no indication of how early in life Monahan developed his interest in natural history, or what guiding influences may have fostered his fascination with birds and wildflowers. But during his college career he gravitated toward the study of the natural sciences. Records show that while an undergraduate student at Baker University Monahan worked as an assistant in the Biology Department. Graduating at age 29 in 1905, reflects the fact that he worked his way through those college years, no doubt adding to the length of his undergraduate tenure. Monahan Jr. remembers his dad reminiscing of those college years, working both in a bank and on a farm ('shucking 100 bushels of corn a day'). Nevertheless, immediately upon graduation from Baker in 1905, Monahan was assigned to the Chair of Systematic Zoology where he taught as a professor of biology and taxidermy.

That the newly graduated Monahan was kept on staff and

promoted into the capacity of a teaching professor and curator of the college museum, speaks highly of his reputation, abilities and academic promise. Moreover, his stipend-funded leave of absence to undertake the epic 1907 collecting expedition to the wilds of Idaho resounds of further institutional confidence in him. Recall that this financial commitment came during one of the most ominously fragile economies of our nation's history (similar to our current 2008-2010 financial collapse). In the face of this widespread financial gloom, Professor Monahan garnered complete administrative approbation.

Through *The Baker Orange*, a campus alumni newsletter, we get snatches of Professor Monahan's activities outside the museum just months prior to his departure for Idaho. For example, in late January of 1907 he represented Baldwin as a delegate to a meeting of the Kansas I.O.O.F.'s in Topeka. In early February he served as a judge in the Biblical Prize Debate, presumably on the Baker campus. In mid February he was back in the field on a wolf hunt with his Baker colleague, Dr. Parmenter. Successful in their wolf quest, Monahan spent the better part of the following week working on the resultant wolf mount for the museum.

Following his return from the Oregon extension of the 1907 Idaho expedition, Monahan labored a full year at Baker preparing his extensive Pacific Northwest collection for the museum. But even then, he took occasional breaks from the museum work to venture into the field to collect additional local specimens. I note that during that 1908-09 year while tirelessly working to prepare his expedition specimens for display, Monahan also served at the campus printing press. I suspect part of that responsibility may have pre-positioned him for the necessary supervision of the printing of hundreds of very technically labeled but professional looking 1 ¾' x 2 7/8" Baker Museum specimen cards. Those labels all bear the name of the museum in fine print and in larger type provides the specimen's identity, with details of date and location of collection. All of the 251 specimens Monahan is credited to have added to the museum that year would have required such a label. It's little wonder he spent time at the printing press.

Though somewhat reclusive in his museum labors that year, Monahan's work did not go unnoticed. The 1909 Baker University Trustee's Minutes record this tribute to his 1908-09 labors on campus: "Professor Monahan having no class work has devoted his whole time to mounting and putting in preparation the material collected during the scientific expedition of 1907-'08. Too much praise and credit cannot be given him for his indefatigable and enthusiastic work throughout the year. And many times he has labored under a disadvantage, and in addition to his own work, conducted the press."

Clearly, Monahan's artful labor on his expedition collection was steadily and noticeably embellishing the Baker Museum. The university's June 9, 1909 Museum Report summed it up like this: "This year has been one of unparalleled prosperity, and no time in the history of Baker University has the museum been so well equipped, so thoroughly classified..." Along with the burgeoning number of new specimens added to the museum that year was the purchase of a number of new display cases to properly house them.

By the 1909-10 school year Monahan was thrust back into the mainstream of campus responsibilities, which were legion. In addition to his ongoing museum duties and teaching responsibilities both in collage classes and at the Baker Academy, Monahan served on several campus committees. The 1909 Baker Catalogue lists him on committees for Contests in Debate, Oratory and Declamation; Literary Society; Campus Life; Publications; and the 50th Anniversary Committee. Every college professor knows the drudgery of committee work; and that committee load strikes me as formidable.

Between 1908 and 1910, *The Baker Orange* alumni newsletter occasionally reported on some of Monahan's excursions and specimen acquisitions. Those early alumni records openly credit him as being a major contributor to the success of the Baker Museum, second in importance only to Dr. Charles Sylvester Parmenter. That was indeed high praise, considering that Dr. Parmenter labored on for many years and became such a revered figure on campus that he was later honored as the namesake for one of the Baker University Science Buildings. A 1929 campus newsletter reported that between 1905 and 1910 Monahan added to the museum an estimated 1500 animal specimens, including birds and insects, as well as 1000 plant specimens. Surely, the bulk of that bounty came from his 1907 Idaho trek.

In June of 1910 Monahan's tenure with Baker University came to an end. There is no reason offered anywhere for this decision to move on; only that he began studies in Comparative Zoology that summer at the University of Chicago. Registration documents show that during that time Monahan was under contract to work 80 hours a term in the Zoology Department for a total of $120.00 to be applied to his tuition. Not much else is known about this transitional experience. Nor do we know why a year later Monahan switched his academic pursuits to that of Medical School at the University of Kansas. Monahan Jr. remembers his dad simply stating that he "ran out of money." Still, medicine seems a far cry from his long dedicated run in natural history. Of course, history is replete with the stories of famous naturalists of old who became medical doctors out of necessity to support their habit of natural history study.

The next several years found Monahan setting a more permanent course for his life. On June 10, 1914, he received his Doctor of Medicine Degree from the University of Kansas and began his practice in Kansas City as a general Practitioner and Obstetrician. While in residency at the University of Kansas Medical School Hospital, he met and fell in love with an R.N. named Jessie E. Foster. Jessie, the daughter of a well-to-do lawyer in Sedgwick, Kansas, would have been about 21 years of age at the time. Monahan, on the other hand, was pushing forty. Through some mild and well-meaning subterfuge Monahan led his sweetheart to believe that he was a younger man than his actual years. Obviously, the ruse, and his good looks and charm won out because the couple would eventually marry.

About that time World War I erupted onto the world scene and in 1916 Monahan enlisted in the Army Medical Corps. The young doctor never left the states but instead, was stationed on the home front at Ft. Sill, Oklahoma. There he engaged in a battle as deadly as any combat zone on the allied front. The enemy he fought was not the Kaiser's German forces, but the Spanish Influenza, one of the most deadly viruses the world has yet experienced. That world-wide pandemic eventually took its toll of 50 million lives, far exceeding the battlefield carnage of W.W. I. It is ironic, how the insane and intentional splattering of bullets, bombs, bayonets, and mustard gas couldn't inflict nearly as much human death and tragedy as the silent submicroscopic killer virus. Today, readers living in the first decade of the twenty-first century can relate to this sort of passive world-wide flu outbreak because as I write this book the world is combating another such pandemic – the HINI Influenza. Providentially, thus far this virus is manifesting itself in a far less deadly form than the 1918 Spanish Influenza.

During his entrenchment in battling the deadly flu pandemic, Monahan himself contracted the virus, surviving the onslaught. On June 18, 1918 he and Jessie were married. In October of 1919, Monahan's only offspring, a son, Elmer Perry Monahan Jr., was born. Now Monahan the pioneer naturalist, former college professor and museum curator, medical doctor and father had someone with whom he could share his love for the outdoors and his intrigue with nature.

Even in his busy medical practice, Monahan never drifted too far away from his love of natural history. While Elmer P. Jr. grew into boyhood, father and son enjoyed close involvement in Boy Scouts of America. This provided a perfect outlet for Monahan to share his interest in natural history and his experience in nature lore. For many years he was guest speaker at local Boy Scout troops, often times demonstrating his rare taxidermy skills. To skin and complete a bird taxidermy mount took Monahan but an hour! It comes as no surprise that Monahan's son, Elmer P. Jr. was very active in scouting, and eventually attained its highest rank of Eagle Scout.

For a number of years, even after his son was born, Monahan maintained some connection with Baker University through his service on one or two campus committees. Elmer Monahan Jr. recalls fond memories of accompanying his dad to some of those meetings where he played among the stuffed animals of the museum while his father conducted his business. Even then, in the mid- to late-1920's the museum was showing signs of deterioration. Eventually, the demands of medical practice and family obligations dissolved Monahan's close ties to Baker.

I can't help but wonder what effect the sight of the moribund museum had on Monahan's disassociation with Baker University. Next to Dr. Parmenter, Monahan was openly acclaimed as the major contributor to the Museum's once premier reputation. To a large extent, the Baker Museum's history was *his story.* Witnessing the deterioration of the museum and its slide from campus prominence and pride must have been difficult.

In 1929 Monahan embarked on yet another daring quest. This one – with flashes of his dauntless yen for adventure clearly evidenced back in 1907 – spanned the globe. This trek was a medical expedition of sorts. After spending several months teaching himself German, he took a year off from his Kansas City practice and crossed the Atlantic to study dermatology in Vienna, Austria. At the time Vienna was the world center of skin research and teaching. Officially, it is said that the reason for this sharp turn in specialization was because Monahan always had a keen interest in skin diseases. That harkens us back in time to his 1907 bodacious bout with the ravages of poison oak over his face and neck. Could that affliction have been one of the early inspirations for later delving into the science of skin diseases?

These many years later, Monahan's son, Elmer P. Jr., confided to me a more practical explanation for this switch to Dermatology: his dad had simply grown weary of mid-night calls to deliver babies, or tend to other medical emergencies. Certainly, a Dermatologist faces fewer emergency cases and enjoys a normal work schedule. Considering that by 1929 Monahan was 55 years old, one can give some credence to this line of reasoning.

It is revealing but perhaps not surprising that even while immersed in his specialized medical studies in Austria, Monahan maintained his abiding interest in birds. In the heartland of the renowned and eventual Nobel Prize winning bird behaviorist, Konrad Lorenz, how could he not? Whether while traveling to and from his studies, or by occasional field excursions across the Austrian landscape, Monahan learned the birds he saw, and found such joy in them he conceived the idea of writing a book entitled: *Birds Along the Auto Trail.* His keen naturalist eyes which 22 years earlier plumbed bird-life along dusty wagon trails of Idaho, now posted similar observations from a more modern conveyance, the automobile. Of course, this time his focus centered on his binoculars instead of down the length of a gun barrel.

Through his abiding interest in birds, somewhere along the way Monahan struck up an acquaintance with one of North America's premier ornithologists, Alexander Wetmore. I would love to know how this friendship came about. All I know is that Wetmore was an undergraduate at Kansas University in 1911-12, overlapping the time Monahan was at the same institution in Medical School. In the beginning, Wetmore was planning to study medicine, but later switched to ornithology and paleontology. Could these two men have met in science classes? Might they have bumped into each other in the Museum area, which I'd bet both frequented from time to time? I can't imagine that Monahan, the former ace expedition collector, taxidermist, and museum curator didn't occasionally find his way into the Museum sector. Perhaps even before Medical School, while Monahan labored indefatigably in the Baker Museum, he and Wetmore might have met through correspondence about museum specimens. Then again, they may have met through Monahan's long membership in the American Ornithologists' Union (A.O.U.).

Wetmore went on to a stellar career in the Smithsonian Institute, serving first as Assistant Secretary in 1925 and later promoted to Secretary in 1945. During that time he was very active in avian research, with both fossil and living birds. Interestingly, what most people don't know about Wetmore is his distinction of being one of the 27 members – perhaps the youngest, at 14 years of age – of Frank M. Chapman's first-ever Christmas Bird Count, on Christmas day 1900.

For many years Monahan treasured one of his choice bird books, co-authored by Wetmore: the 1937 National Geographic Society two-volume, *The Book of Birds*. Monahan Jr. remembers his dad's love for those books and his frequent and admiring references to his friend Alexander Wetmore but how and when this friendship came about remains another lingering mystery.

Ironically, Alexander Wetmore left his footprints in Idaho and on my own life. Between 1929 and 1931 he was the first avian paleontologist to visit the Hagerman Fossil Beds National Monument in southwest Idaho. Many years later in 1969 he confirmed my identification of a very rare blue grosbeak that a friend photographed near my hometown of Glenns Ferry, Idaho. Regrettably, at the time I had no inkling of his recently deceased friend, Elmer Perry Monahan. Nor did I know that a mere four years later I would discover his friend's cryptic 1907 diary story which would put me on a three-decade quest for its lost biological bullion.

In 1949 Monahan was elected an Associate of the American Ornithologists' Union, another testimony to his active and enduring interest in birds. Despite this long association with these professional ornithologists there is no indication that he ever published in any of the more technical ornithological journals. Occasional articles for newspapers and popular magazines seemed to have been his usual outlet for writing about birds. This is not to say that he did not write professionally. During his many years as a teaching member of the Department of Dermatology, at the University of Kansas School of Medicine we know that he occasionally published within journals of his profession. One known example is a several page piece he authored in the *Research Clinic Bulletin* (Vol. 2: no.7, 1951), entitled "Acne Vulgaris – Nemesis of Youth".

In the 1930's after his return from Austria, Monahan acquired a 6-acrea plot of undeveloped land that encompassed a small year-round creek. On this wild little estate he built a solid rustic cabin complete with a "no nonsense" fireplace. This family acreage long provided him a site for his hobby of raising purebred Hampshire Sheep. Wildflowers still sprinkled the landscape and Monahan was especially proud of a patch of rare Ladys- slipper orchids. His son recalls this home as a place of many picnics and frequent visits by Boy Scouts.

With his changeover from museum collector and curator to the healing arts of medicine, Monahan likewise converted his animal collecting into the more genteel sport of hunting. Until he was in his mid-80's he still joined his son on regular fall deer hunts at a "hunting shack" in the Colorado mountains. As might be expected, Monahan Jr. grew up under the tutelage of his sharp-shooting father. When he was just a young lad, his father presented him with the gift of a Winchester .22 single-shot rifle. In keeping with the single-shot rifle was the sage advice Monahan passed on to his son: "In hunting, the only shot that counts is the first one". The neighborhood rabbit population bore the brunt of the younger Monahan's practice of that instruction.

In his later years Monahan remained active in his medical profession. It wasn't until 1957 at the age of 83 that he finally retired to a woodsy 150 acre plot near Lenexa, Kansas. There he continued his hobby of raising sheep and no doubt enjoyed the wildlings frequenting his acreage. On March 28, 1964, Elmer Perry Monahan died at age 89.

I deeply regret that I never met Elmer Perry Monahan Sr. It would have been possible; I graduated from college the year he died. But by the time I learned of the man – through the epiphany of reading his diary account – it was too late. So I've chased his ghost and his phantom expedition ever since. And this quest has left strong impressions of the man.

It seems to me that Monahan's life – though unpretentious – was filled with courage and character; and a willingness to venture forth toward self discovery. From his humble beginnings of being "farmed out" into host families as a small child, to striving ever upward into the college ranks – uncharted territory for any others of his kin – he grew strong of purpose and with a sense of direction, ever proving himself capable, talented, and engaging with people. In his youth he successfully masterminded and led one of the last pioneer scientific expeditions, over 1200 miles from his college campus. He courageously battled a deadly pandemic while serving his country and society, surviving his own infection. His intrepid educational journey to the hallowed medical hallways of Vienna – late in life, and during the Great Depression – showed not only a quiet self-confidence but professional verve. He never abandoned his deep naturalistic passion for the lilies of the field and the birds of the air. He lived long, he lived well, and through his abiding love of observing nature, he lived in the moment.

Springbeauty

·18·

A PLACE OF HONOR

With this book, the long hidden and nearly forgotten has finally been lifted from dusty oblivion back into the light of history. Monahan, the 1907 Baker University biological expedition, and the vanished Baker University Museum of Natural History have been resurrected into modern memory. Tattered, newly discovered 103-year-old field journals are opened to unveil long-held secretes of early Idaho-Oregon pioneer biotic samplings. Several century-old bird specimens are rescued and now stand in stolid testimony to a lost legacy. Will all this remain forever bound between the covers of this book? Or is there a place of honor at some commemorative site where these tangible proofs might yet be displayed along with their story to forever remind people of their remarkable history?

In recent communication with my valuable contact and friend in the Baker University Biology Department, Dr. Roger Boyd, he suggested a probable and honorable home for the old museum specimens and other 1907 Expedition remnants. Baker University owns a 573 acre wetlands area which the father-son Boyd biologists have managed since 1968. Over the years the Kansas Department of Transportation (KDOT) has found need of constructing a bypass over the edge of the wetlands, and as mitigation, brokered a several million-dollar deal with Baker University to clinch the project. In addition to receiving an additional 400 acres, Baker University also garnered a million dollars for restoration of the Baker Wetlands which is currently underway. As soon as the economy rebounds, Phase 2 of the KDOT-Baker University deal will kick in another

1.2 million dollars to build a Visitor Center complete with a 4.5 million dollar endowment to staff and manage the complex.

Once the Baker Wetlands Visitor Center is in place, Dr. Boyd envisions it housing attractive museum displays of area wildlife specimens, including a special place to highlight the old Baker Museum of Natural History and the 1907 Monahan Expedition. When this ambitious dream of a wetlands facility comes to pass, the glory days of Monahan's era of museum work and his remarkable Northwest odyssey will finally be enshrined for posterity, and no longer forgotten to the ages.

Clarkia

ACKNOWLEDGEMENTS

Much is owed to Dr. Homer E. Socolofsky for his initial deciphering of Monahan's original diary and researching the circumstances surrounding this expedition. I am indebted also for his liberal loan of some of his research notes, and generous offer of insights into the history of this fascinating bit of Idaho history. Several Baker University personnel provided critical help in ferreting out details of Monahan's role at the university and his contributions to the old museum. In 1976 Dr. V.E. Nelson was extremely helpful in salvaging the few surviving museum specimens from the 1907 expedition. At that same time he, and later in 1991. Dr. J. Keith Keeling made efforts to locate any museum records or any other campus documentation detailing the specimen acquisitions of Monahan's Idaho expedition. Mrs. Maxine Kreutizger, longtime Baker University Archivist, assisted in searching for campus documents of the 1907 Monahan expedition. Current Baker Archivist Jen McCollough was very helpful in providing old photographs of Monahan, Ray Green, and various stages of the old museum..

Thirty-six years after failing to locate any written records of Monahan's Idaho collections, an extraordinary breakthrough came in August of 2009. Through the attentiveness of Dr. Roger Boyd of the Baker University Biology Department, aided perhaps by some providential past efforts of his father, Dr. I. L. Boyd, Roger found and brought to my attention Monahan's long missing 1907 original field journals! The contents of those journals helped to put scientific meat on the bare bones scant records of the 1971 published diary.

I am deeply grateful to Elmer Perry Monahan Jr. for his generous provision of personal memories, family history notes and memorabilia relating to his father. Without that infusion of his insight and family archival information, my portrayal of his dad's full and interesting life would have been severely wanting. And certainly, Monahan Jr.'s unsolicited and totally unexpected financial contribution to this book's publication was a boost to its completion. Above all, I want to acknowledge my appreciation for the warm and delightful friendship I now enjoy with this son of the 1907 expedition leader.

I could not have retraced this expedition with my "student naturalists" without the help of two people. Allen R. Dale and Timothy H. Craig contributed immeasurably with their instructional duties in the several field biology classes that studied and retraced the expedition. Beyond their biological expertise and teaching prowess, they both offered great encouragement, and infused enthusiasm into the classes.

The Northwest Nazarene University Science Math Alumni Association contributed financial support for my initial planning and preparatory scouting out of portions of the 1907 expedition route. Coral Mattei and LaRita Shandorff of the Technical Services at the NNU Riley Library kindly assisted me in tracking down early 1900's bird field guides for reference as well as locating the May 25, 1907 *Idaho Statesman* Newspaper article about the Baker Expedition.

I am indebted to my son Shane R. Powers and my sister-in-law Roberta Thompson for their photographic expertise in documenting some of the 1907 specimens and field journals for purposes of this book. This book has benefitted enormously from suggestions and editing by my son Ryan S. Powers, my friend Laurene Stanford, and my wife, Willa. My sincere thanks to Dr. Charles H. Trost for not only his mutual interest in this project at its very onset, but also for his careful scrutiny of the avian data recorded herein. I also thank Ken Kauffman for his careful reading of the galley proof of this book. His discerning eye caught what we hope was the last of editorial needs.

My friend and former student, Peter Gaede, provided the cover art for this book, and another friend and back-packing buddy, Paul Kinsman, added his artistic touch and guidance to cover design, book graphics and page set-up for the printing process.

LITERATURE CITED

Burleigh, Thomas D. 1972. *Birds of Idaho,* The Caxton's Printers, LTD., Caldwell, Idaho.

Callandar, Sine. 1987. *Early Days in Long Valley.* Idaho Historical Society, Boise, Idaho.

Coues, Elliott. 1903. *Key to North American Birds Vol. 1 & 2,* Dana Estes and Company, Boston

Earle, A. Scott with Jane Lundin. 2008. *Idaho Mountain Wildflowers: A Photographic Compendium* (2nd. Edition), Larkspur Books, Boise, Idaho.

Evens, Howard Ensign. 1993. *Pioneer Naturalists,* Henry Holt and Company, New York.

Gallencamp, Charles. 2001. *Dragon Hunter,* Penguin Group, New York, New York.

Ingram, Beverly. 1992. *Looking Back: Sketches of Early Days in Idaho's High Country,* Maverick Publications, Inc. Bend, Oregon.

National Audubon Society's *The Sibley Guide to Birds.* 2001. Alfred A. Knoph, New York.

Pottenger, Cynthia. *My Roseberry.* McCall, Idaho: *The Star News,* 1972.

Powers, Leon R. and Charles H. Trost. 1977. Early Records of Avifauna from a 1907 Scientific Expedition into Idaho, *The Murrelet* 58 (3): 87-88.

Reed, Chester. 1906. *Bird Guide: Land Birds East of the Rockies and Water Birds, Game Birds, and Birds of Prey*, Doubleday, Doran & Company, New York.

Socolofsky, Homer. 1971. *Idaho Yesterdays*. Winter,1970-71. pp.18 -35.

The Idaho Statesman, May 25, 1907, Boise, Idaho.

The Idaho Statesman, February 10, 2008. Boise, Idaho.

The Long Valley Advocate, September 21, 1905. Lardo, Boise County, Idaho.

The Long Valley Advocate, September 5, 1907. Lardo, Boise County, Idaho.

The Long Valley Advocate, September 19, 1907. Lardo, Boise County, Idaho.

Vizgirdas, Eay S. 2003. *Useful Plants of Idaho*. Idaho State University Press, Pocatello, Idaho.

Weidensaul, Scott. 2007. *Of a Feather: A Brief History of American Birding,* Harcourt, Inc., Orlando.

Welty, Joel Carl. 1988. *The Life of Birds,* Saunders College Publishing, San Francisco, California.

Woods, Sheldon, Editor. 2002. *Valley County, Idaho – Prehistory to 1920,* Action Publishing, Donnelly, Idaho.

ORDER FORM

Name ____

Address ______________________________________

City/State/Zip ________________________________

Phone _______________________________________

Enclosed is my Check for:

___THE FORGOTTEN EXPEDITION

($16.95, includes tax, + $3.50 for shipping and handling = $20.45)

___ DEAD OWLS FLYING

($15.00, includes tax, + $3.50 for shipping and handling = $18.50)

___ A HAWK IN THE SUN (limited supply)

(Reduced price: $12.00, includes tax, + $3.50 for ship/hand = $15.50)

TOTAL enclosed:______________________

Please send Order Form and payment to:

Look-About Books

Box 1907

Nampa, ID 83653 – 1907

Phone: (208) 466-6260

E-mail: lookaboutbooks@leonpowers.com

Website: leonpowers.com